Characterizing Natural Organic Mat

Water Treatment Processes aı.

T0231381

CHARACTERIZING NATURAL ORGANIC MATTER IN DRINKING WATER TREATMENT PROCESSES AND TRAINS

DISSERTATION

Submitted in fulfilment of the requirements of

the Board for Doctorates of Delft University of Technology

and of the Academic Board of the UNESCO-IHE

Institute for Water Education

for the Degree of DOCTOR

to be defended in public on

Thursday 15 November, 2012, at 12:30 hours

in Delft, the Netherlands

By

Saeed Abdallah BAGHOTH

Master of Science in Sanitary Engineering
UNESCO-IHE Institute for Water Education, The Netherlands

born in Kamuli, Uganda

CRC Press
Taylor & Francis Group
Boca Raton London New York

CRC Press is an imprint of the
Taylor & Francis Group, an **informa** business
A BALKEMA BOOK

This dissertation has been approved by the supervisor:
Prof. dr. G.L. Amy

Composition of Doctoral Committee:

Chairman	Rector Magnificus Delft University of Technology
Vice-chairman	Rector UNESCO-IHE
Prof. dr. G.L. Amy	UNESCO-IHE/ Delft University of Technology, Supervisor
Prof. dr. L.C. Rietveld	Delft University of Technology
Prof. dr. T. Leiknes	Norwegian University of Science and Technology, Trondheim, Norway
Prof. dr. Ing. M. Jekel	Technical University of Berlin, Berlin, Germany
Dr. S.K. Sharma	UNESCO-IHE
Dr. M. Dignum	Waternet, The Netherlands
Prof. dr. M.D. Kennedy	UNESCO-IHE/ Delft University of Technology, reserve member

CRC Press/Balkema is an imprint of the Taylor & Francis Group, an informa business

© 2012, Saeed Abdallah Baghoth

All rights reserved. No part of this publication or the information contained herein may be reproduced, stored in a retrieval system, or transmitted in any form or by any means, electronic, mechanical, by photocopying, recording or otherwise, without written prior permission from the publishers.

Although all care is taken to ensure the integrity and quality of this publication and the information herein, no responsibility is assumed by the publishers nor the author for any damage to the property or persons as a result of operation or use of this publication and/or the information contained herein.

Published by:
CRC Press/Balkema
PO Box 447, 2300 AK Leiden, the Netherlands
e-mail: Pub.NL@taylorandfrancis.com
www.crcpress.com - www.taylorandfrancis.co.uk - www.ba.balkema.nl

ISBN 978-1-138-00026-1 (Taylor & Francis Group)

Dedication

This thesis is dedicated to my wife and children, without whose patience and perseverance I would not have been able to complete.

Acknowledgements

I would like to thank Prof. Gary Amy for accepting to be my Promoter and for sharing with me his expert knowledge. I learnt a lot from your experience it is with your guidance and encouragement that I was able to progress to the end of my PhD research. I would also like to extend my sincere appreciation to my supervisor, Dr. Saroj K. Sharma, who kept me focused on my goal and whose guidance played an important part in the finalization of my thesis. Your useful comments contributed immensely towards honing my writing skills and for that I am particularly grateful.

I would like to thank SenterNovem agency of the Dutch Ministry of Economic Affairs for partly funding my PhD study through the collaborative IS NOM research project involving KWR water research institute, UNESCO-IHE Institute for Water Education, Delft University of Technology and the water supply companies of Vitens and Waternet, both of The Netherlands. I am grateful to all the members of the IS NOM research group for the collaboration and exchange of useful ideas. I am particularly thankful to Anke Grefte, PhD researcher with Delft University of Technology, for her help in collecting samples from Waternet and for allowing me to conduct F-EEM analysis of samples from her pilot plant set-up. I am grateful to Waternet for providing me with the invaluable SEC-OCD data for most of the samples I collected from their treatment plants.

I would like to acknowledge the financial support of Syndicat des Eaux d'lle de France (SEDIF), France, which allowed me to carryout out part of the research involving water treatment plants for the Paris suburbs. I am grateful to of Poitiers University for the hospitality during the two months I worked in their laboratory.

I am very grateful to Julius K. Mwesigwa, Mariano R. Tesoura and Mosebolatan K. Bola for their contribution with the experimental work during their MSc.study at UNESCO-IHE students. I would like to thank my colleagues Sung Kyu Maeng (Dr.), S.G. Salinas Rodriguez (Dr.) for the close collaboration during our PhD research. I am grateful to my other PhD colleagues Tarek Waly (Dr.), Chol Abel, Abdulai Salifu, Mohamed Babu (Dr.), Valentine Uwamariya and Loreen O. Villacorte who made me forget the loneliness of being away from my family. To the UNESCO-IHE laboratory staff Fred Kruis, Frank Wiegmen, Peter Heerings, Don van Galen and Lyzette Robbemont, I am very grateful for your help in the laboratory.

I am very grateful to my employer, Kamuli District Local Government, Uganda, for offering me the study leave to pursue the PhD study. The support extended to me meant a lot to the wellbeing of my family back home. Kamuli has been, and will always, be in my heart!

Lastly, I would like to express my heartfelt gratitude to members of my family for their patience and what they had to endure during my long period of stay away from them. I salute all of you and pray that the almighty rewards you bountifully! For those that I have not been able to individually acknowledge, your contributions during my PhD study, in whatever form, were very helpful and I am sincerely thankful.

Abstract

Over the last 10–20 years, increasing natural organic matter (NOM) concentration levels in water sources have been observed in many countries due to issues such as global warming, changes in soil acidification, increased drought severity and more intensive precipitation events. In addition to the trend towards increasing NOM concentration, the character of NOM can vary with source and time (season). The great seasonal variability and the trend towards elevated NOM concentration levels impose challenges to the drinking water industry and water treatment facilities in terms of operational optimization and proper process control. By systematic characterization, the problematic NOM fractions can be targeted for removal and transformation. Therefore, proper characterization of the NOM in raw water or after different treatment steps would be an important basis for the selection of water treatment processes, monitoring of the performance of different treatment steps, and assessing distribution system water quality.

NOM is a heterogeneous mixture of naturally occurring organic compounds found abundantly in natural waters and originates from living and dead plants, animals and microorganisms, and from the degradation products of these sources. NOM in general significantly influences water treatment processes such as coagulation, oxidation, adsorption, and membrane filtration. In addition to aesthetic problems such as colour, taste and odour, NOM also contributes to the fouling of membranes, serves as precursor for the formation of disinfection by-products (DBPs) of health concern during disinfection/oxidation processes and increases the exhaustion and usage rate of activated carbon. Furthermore, the biodegradable fraction of NOM may promote microbial growth in water distribution networks. The efficiency of drinking water treatment is affected by both the amount and composition of NOM. Therefore, a better understanding of the physical and chemical properties of the various components of NOM would contribute greatly towards optimization of the design and operation of drinking water treatment processes.

Because it may contain thousands of different chemical constituents, it is not practical to characterize NOM on the basis of individual compounds. It is more feasible and the general practice to characterize it according to chemical groups having similar properties. These groups are commonly isolated by methods which involve concentration and fractionation of bulk NOM. However, these methods are often laborious, time consuming and may involve extensive pre-treatment of samples which could modify the NOM character. They are also difficult to install for online measurement and are not commonly used for monitoring of NOM in drinking water treatment plants.

Analytical techniques that can be used to characterize bulk NOM without fractionation and pre-concentration and with minimal sample preparation are becoming increasingly popular. High performance size exclusion chromatography (HPSEC) and fluorescence excitation-emission matrix (F-EEM) spectroscopy are increasingly used for NOM characterization in drinking water. More detailed information about NOM can be obtained by using F-EEM spectra and parallel factor analysis (PARAFAC), a statistical method used to decompose multi-dimensional data.

The aim of this research was to contribute to a better understanding of the character of NOM before and after treatment by different drinking water treatment processes as well as in the water distribution network using multiple NOM characterisation tools like F-EEM, SEC with ultraviolet absorbance (UVA) and dissolved organic carbon (DOC) detectors (SEC-OCD), and other bulk NOM water qualities such as UVA at 254 nm (UVA_{254}), specific UVA_{254} (SUVA) and DOC. These complementary techniques can provide information on the fate of NOM fractions that negatively impact treatment efficiency, promote biological re-growth in water distribution systems and provide precursors for DBPs in systems that use oxidation/disinfection processes. It is expected that this would permit the optimization of NOM removal during water treatment in terms of quantity as well as specific NOM fractions of operational and health concern.

NOM in water samples from two drinking water treatment trains with distinct water quality, and from a common distribution network with no chlorine residual, was characterized and the relation between biological stability of drinking water and NOM was investigated through measurements of assimilable organic carbon (AOC). NOM was characterised according to F-EEM, SEC-OCD and AOC. The treatment train with higher concentrations of humic substances produced more AOC after ozonation. NOM fractions determined by SEC-OCD, as well as AOC fractions, NOX and P17, were significantly lower for finished water of one of the treatment trains. F-EEM analysis showed a significantly lower humic-like fluorescence for that plant, but no significant differences for tyrosine- and tryptophan-like fluorescence. For all of the SEC-OCD fractions, the concentrations in the distribution system were not significantly different than in the finished waters. For the common distribution network, distribution points supplied with finished water containing higher AOC and humic substances concentrations had higher concentrations of adenosine triphosphate (ATP) and *Aeromonas* sp. The number of aeromonads in the distribution network was significantly higher than in the finished waters, whereas the total ATP level remained constant, indicating no overall bacterial growth.

The use of F-EEMs and PARAFAC to characterize NOM in drinking water treatment and the relationship between the extracted PARAFAC components and the corresponding SEC-OCD fractions was investigated. A seven component PARAFAC model was developed and validated using 147 F-EEMs of water samples from two full-scale water treatment plants. Five of these components are humic-like with a terrestrial, anthropogenic or marine origin, while two are protein-like with fluorescence spectra similar to those of tryptophan-like and tyrosine-like fluorophores. A correlation analysis was carried out for samples of one treatment plant between the maximum fluorescence intensity (F_{max}) of the seven PARAFAC components and the NOM fractions of the same samples obtained using SEC-OCD. The sample DOC concentrations, UVA_{254}, and F_{max} for the seven PARAFAC components correlated significantly ($p < 0.01$) with the concentrations of the SEC-OCD fractions. Three of the humic-like components showed slightly better predictions of DOC and humic fraction concentrations than did UVA_{254}. Tryptophan-like and tyrosine-like components correlated positively with the biopolymer fraction. These results demonstrate that fluorescent components extracted from F-EEMs using PARAFAC could be related to previously defined NOM fractions and could provide an alternative tool for evaluating the removal of NOM fractions of interest during water treatment.

NOM in water samples from two drinking water treatment trains was characterized using SEC-OCD and F-EEMs with PARAFAC. These characterization methods indicated that the raw and treated waters were dominated by humic substances. The PARAFAC components and SEC-OCD fractions were then used to evaluate the performance of the treatment plants in terms of the removal of different NOM fractions. Whereas the coagulation process for both plants may be optimized for the removal of bulk DOC, it is not likewise optimized for the removal of specific NOM fractions. A five component PARAFAC model was developed for the F-EEMs, three of which are humic-like, while two are protein-like. These PARAFAC components and the SEC-OCD fractions proved useful as additional tools for the performance evaluation of the two water treatment plants in terms of the removal of specific NOM fractions.

The impact of different water treatment processes for removal of NOM in surface and ground waters on the fluorescence characteristics of the NOM was investigated. The study focuses on the fluorescence spectral shifts of a humic-like peak (peak C), at an excitation wavelength in the visible region of 300-370 nm and an emission wavelength between 400 and 500 nm, and investigates the amount of error in the determination of the fluorescence intensity maximum if the shift in the location of peak C is not taken into account. Coagulation of surface and ground water with iron chloride and alum resulted in a shift in the emission wavelength of humic-like peak C of between 8 and 18 nm, and an error in the maximum fluorescence intensity ranging between 2% and 6% if the shift is not taken into account. There was no significant difference in the spectral shift of peak C or in the error in the maximum fluorescence intensity between coagulation alone and coagulation followed by ozonation of ground water. NOM removal with ion exchange (IEX) alone generally resulted in a higher shift in peak C and a higher percentage error in the maximum fluorescence intensity than with coagulation, biological activated carbon (BAC) filtration or a combination of treatments. The impact of IEX treatment on the error of maximum fluorescence intensity was higher for surface than for ground waters, likely due to differences in molecular weight distribution of surface and ground water NOM. The results demonstrate that for NOM removal treatments other than IEX, the errors in the maximum fluorescence intensity that would result from ignoring the fluorescence spectral shifts are generally low ($\leq$ 5%), and a fixed excitation emission wavelength pair for peak C could be used for online monitoring of NOM in water treatment plants.

Incorporation of F-EEMs to improve the monitoring of concentrations of DOC and total trihalomethanes (THMs) in drinking water treatment was evaluated. Predictive models were developed for the removal of NOM and the formation of THMs after chlorine disinfection in a full-scale drinking water treatment plant using several measured water quality parameters. Whereas the use of PARAFAC fluorescence components slightly improved the prediction of finished water DOC concentration, the prediction accuracy was generally low for both simple linear and multiple linear regressions. The applied coagulation dose could be predicted ($r^2 = 0.91$, $p < 0.001$) using multiple linear regressions involving temperature, UVA_{254}, total alkalinity, turbidity and tryptophan-like fluorescence (peak T). The total THMs concentration of the finished water could be predicted ($r^2 = 0.88$, $p < 0.001$) using temperature, turbidity, ozone dose, UVA_{254}, fluorescence peak T and a humic-like peak (peak M), with an excitation maximum at 310 nm and an emission maximum at 410 nm.

This research contributes to our knowledge of the character of NOM and the impact of different drinking water treatment processes on its characteristics. It demonstrates the potential of using multiple NOM characterization tools for the selection, operation and monitoring of the performance of different water treatment processes and the assessment of the water quality in a water distribution system.

Table of contents

Chapter 1

INTRODUCTION

1.1   Background

Over the last 10–20 years, increasing natural organic matter (NOM) concentration levels in water sources have been observed in many countries due to issues such as global warming, changes in soil acidification, increased drought severity and more intensive rain events (Fabris et al., 2008). In addition to the trend towards increasing NOM concentration, the character of NOM can vary with source and time (season). The great seasonal variability and the trend towards elevated NOM concentration levels impose challenges to the water industry and water treatment facilities in terms of operational optimization and proper process control (Fabris et al., 2008). By systematic characterization, the problematic NOM fractions can be targeted for removal and transformation. Therefore, proper characterization of the NOM in raw water, or after different treatment steps, would be an important basis for the selection of water treatment processes, monitoring of the performance of different treatment steps, and assessing distribution system water quality.

NOM is a heterogeneous mixture of naturally occurring organic compounds found abundantly in natural waters and originates from living and dead plants, animals and microorganisms, and from the degradation products of these sources (Chow et al., 1999). Its chemical character depends on its precursor materials and the biogeochemical transformations it has undergone (Aiken and Cotsaris, 1995). Its concentration, composition and chemistry are highly variable and depend on the physicochemical properties of the water such as temperature, ionic strength and pH and the main cation components present; the surface chemistry of sediment sorbents that act as solubility control; and the presence of photolytic and microbiological degradation processes (Leenheer and Croue, 2003).

NOM in general significantly influences water treatment processes such as coagulation, oxidation, adsorption, and membrane filtration (Lee et al., 2006). Some NOM constituents are particularly problematic. In addition to aesthetic problems such as colour, taste and odour, NOM also contributes to the fouling of membranes, serves as precursor for the formation of disinfection by-products (DBPs) of health concern during disinfection/oxidation processes (Owen et al., 1998) and increases the exhaustion and usage rate of activation carbon. Furthermore, The biodegradable fraction of NOM is a carbon source for bacteria and other microorganism and may promote microbial growth and corrosion in the water distribution networks (van der Kooij, 2003; Amy, 1994; Owen et al., 1993). Thus, in order to minimise these undesirable effects, it is essential to limit the concentration of NOM in the treated water. However, the efficiency of drinking water treatment is affected by both the amount and composition of NOM. Therefore, a better understanding of the physical and chemical properties of the various components of NOM would contribute greatly towards optimization of the design and operation of drinking water treatment processes.

Many studies and reviews have been undertaken on the structural characterization of aquatic NOM (Frimmel, 1998; Abbt-Braun et al., 2004; Leenheer, 2004) but its structure and fate in drinking water treatment (individual processes and process trains) are still not fully understood. Because it may contain thousands of different chemical constituents, it is not practical to characterize NOM on the basis of individual compounds. It is more feasible and the general practice to characterize it according to chemical groups having similar properties. These groups are commonly isolated by methods which involve concentration and fractionation of bulk NOM (Frimmel and Abbt-Braun, 1999; Peuravuori et al., 2002). Whereas these methods provide valuable insight into the nature of NOM from diverse aquatic environments, they are often laborious, time consuming and may involve extensive pre-treatment of samples which could modify the NOM character. They are also difficult to install for online measurement and are not commonly used for monitoring of NOM in drinking water treatment plants.

Analytical techniques that can be used to characterize bulk NOM without fractionation and pre-concentration and with minimal sample preparation are becoming increasingly popular. Non-destructive spectroscopic measurements require small sample volumes, are simple in practical application and do not require extensive sample preparation. These techniques are widely used for qualitative and quantitative characterization of NOM (Leenheer et al., 2000; Senesi et al., 1989). Ultraviolet (UV) absorbance, which is typically measured at a wavelength of 254 nm (UVA_{254}), is commonly used as a surrogate measure of the NOM concentration present in natural and treated waters. However, one drawback of UVA_{254} measurements is that bulk NOM as well as NOM fractions typically exhibit nearly featureless absorption spectra, showing decreasing absorbance with increasing absorbance wavelength (Korshin et al., 2009 ; Hwang et al., 2002). The lack of peaks is attributed to overlapping absorption bands of a mixture of organic compounds in NOM and to the complex interactions between different chromophores (Chen et al., 2002). UVA_{254} correlates with the hydrophobic fraction of NOM and its use may underestimate the dissolved organic carbon (DOC) concentration of water samples with less aromatic NOM. Nevertheless, UVA_{254} is a useful tool in drinking water treatment practice for on-line monitoring of DOC concentrations (Edzwald et al., 1985; Amy et al., 1987). Specific UV absorbance (SUVA), which is defined as the UVA_{254} of a water sample divided by the DOC concentration, and molar absorptivity at 280 nm have been found to strongly correlate with the aromaticity of a large number of NOM fractions from a variety of aquatic environments (Chin et al., 1994; Weishaar, 2003). SUVA has been used as a surrogate measure of DOC aromaticity (Traina et al., 1990) and as a surrogate parameter to monitor sites for precursors of disinfectant by-products (Croué et al., 2000).

High performance size exclusion chromatography (HPSEC) and fluorescence spectroscopy are two analytical tools that have recently gained popularity for NOM characterization in drinking water. HPSEC separates molecules according to their molecular size or weight and has been widely applied in characterization of NOM in aquatic environments (Chin et al., 1994; Her et al., 2003; Croué, 2004). It has been shown to be very effective in following changes in the NOM distribution along drinking water treatment trains (Vuorio et al., 1998; Matilainen et al., 2002). Fluorescence excitation-emission matrix (F-EEM) spectroscopy, in which repeated emission scans are collected at numerous excitation wavelengths, is a simple, relatively inexpensive and very sensitive tool that requires little or no sample pre-treatment. It has been used to characterize NOM in diverse aquatic environments (Chen et al., 2003; Wu et al., 2003; Coble et al., 1990; Coble et al., 1993; Mopper and Schultz, 1993). More detailed information about NOM character of water samples can be obtained by using F-EEMs and parallel factor analysis (PARAFAC), a statistical method used to decompose multi-dimensional data. F-EEM and PARAFAC have been used in several studies of dissolved organic matter (DOM) in aquatic water samples (Stedmon et al., 2003; Stedmon and Markager, 2005; Hunt and Ohno, 2007; Yamashita and Jaffe, 2008) but have not previously been used in detailed characterization of NOM in drinking water treatment. As well as contributing to a better understanding of NOM, identification of fluorescent components using PARAFAC could be used to track the fate of problematic NOM fractions and to optimise the design and operation of drinking water treatment processes for NOM removal.

HPSEC may be coupled with detectors such as UV, fluorescence or DOC detectors. Significant advancements have been made in the development of size exclusion chromatographic (SEC) separation systems and detectors for the quantification and characterization of varying apparent molecular weight (AMW) NOM fractions (Allpike et al., 2007; Nam and Amy, 2008; Reemtsma et al., 2008; Huber et al., 2011). This research aims at improving our understanding of the character and fate of NOM during different drinking water treatment processes using multiple NOM characterisation tools like F-EEM, SEC with UV and DOC detectors (SEC-OCD) and other bulk NOM water qualities such as UVA_{254}, SUVA and DOC. These complementary techniques could provide information on the fate of NOM fractions that negatively impact treatment efficiency, promote biological re-growth in water distribution systems and provide precursors for DBPs in systems that use oxidation/disinfection processes.

1.2   The need for further research

NOM negatively impacts water treatment processes such as coagulation, oxidation, adsorption, and membrane filtration. It contributes to colour, taste and odour in drinking water and may serve as a precursor for the formation of DBPs. The biodegradable fraction of NOM may promote microbial growth in water distribution networks, particularly in systems which do not maintain a disinfectant residual in the distribution network (van der Kooij, 2003; Amy, 1994; Owen et al., 1993). In order to minimise these undesirable effects, it is essential to limit the concentration of NOM during drinking water treatment. The efficiency of drinking water treatment is affected by both the amount and composition of NOM. Furthermore, the types of DBPs that may be formed during oxidation processes are influenced by the nature of NOM present. However, there is limited knowledge regarding the selection and operation of treatment processes for the removal of specific DBPs precursors rather than of bulk NOM. Biological stability of drinking water, which is the capacity of the water to minimize microbial growth in the distribution system, is influenced by specific fractions of biodegradable organic matter which may be present in very low concentrations. These low molecular weight organics are commonly referred to as assimilable organic carbon (AOC) and may be quantified using bioassay methods. However, the current bioassay methods are not only incapable of detecting and quantifying the full spectrum of microbial growth promoting NOM, but are also laborious and time consuming.

By systematically characterizing NOM, the problematic fractions can be identified and targeted for removal and transformation. Therefore, proper characterization of the NOM in raw water or after different treatment steps would be an important basis for the selection of water treatment processes, monitoring of the performance of different treatment steps, and assessing distribution system water quality.

Many tools that have been used to characterize NOM do not give information about specific NOM fractions while others require sample pre-treatment that are time consuming and labour intensive. Because of its heterogeneity, the structural characterization of NOM is difficult and its structure and fate in drinking water treatment processes and process trains are still not fully understood. NOM is generally characterized according to chemically similar groups which are commonly isolated by methods which involve concentration and fractionation of bulk NOM. Whereas these methods provide valuable insight into the nature of NOM, they are often laborious, time consuming and may involve extensive pre-treatment of samples which could modify the NOM character. They are also difficult to install for online measurement and are, therefore, not commonly used for monitoring NOM in drinking water treatment plants.

This research aims at improving our understanding of the character and fate of NOM during different drinking water treatment processes using multiple NOM characterisation tools such as F-EEM, SEC with UV and DOC detectors (SEC-OCD) and other bulk NOM water qualities such as UVA_{254}, SUVA and DOC. These analytical tools require minimal sample volumes, no pre-treatment and are sensitive. These complementary techniques could provide information on the fate of NOM fractions that negatively impact treatment efficiency, promote biological re-growth in water distribution systems and provide precursors for DBPs in systems that use oxidation/disinfection processes. They could be used to improve the design of water treatment processes and process trains by targeting the removal of specific NOM fractions, resulting in the reduction of DBP formation and chemical and energy use during water treatment. They could also be used for improving process controls of water treatment plants and they offer the possibility for online monitoring of NOM and at low levels of detection which is otherwise not feasible with only DOC or UVA_{254} measurements.

1.3   Objectives of the study

This PhD study was carried out within the context of the IS NOM collaborative research project funded by SenterNovem agency of the Dutch Ministry of Economic Affairs and involved KWR water research institute, UNESCO-IHE Institute for Water Education, Delft University of Technology and the water supply companies of Vitens and Waternet, both of The Netherlands. The goal of the IS NOM project was to improve the biological stability of drinking water through the use of improved treatment technologies for the removal NOM. One of the three PhD studies in the project investigated the improvement of the bioassay methods for the measurement of AOC in drinking water, the second one investigated the use of innovative ion exchange resin treatment for NOM removal and this one focused on the characterization of NOM in drinking water treatment processes and process trains. The aim of this research was to contribute to a better understanding of the character of organic matter in natural waters before and after treatment by different drinking water treatment processes as well as in the water distribution network. It is expected that this would permit the optimization of NOM removal during water treatment in terms of quantity as well as specific NOM fractions of operational and health concern. The specific objectives of this study were:

- To characterize NOM in water samples from source to tap for two water treatment trains in which no chemical residual is applied in the distribution using F-EEM and SEC-OCD.

- To characterize NOM in samples from a drinking water treatment train using F-EEMs and PARAFAC and to investigate the relationship between the extracted PARAFAC components and the corresponding SEC-OCD fractions.

- To use SEC-OCD, F-EEM and PARAFAC to evaluate the performance of different water treatment processes in terms of NOM removal.

- To investigate the shifts in the fluorescence spectra of surface and ground waters during drinking water treatment for NOM removal.

- To investigate the incorporation of fluorescence measurements to improve the monitoring of THM formation in water treatment and to develop predictive models for removal NOM and formation of THMs after chlorine disinfection in drinking water treatment.

1.4   Outline of the thesis

This thesis is organized in eight chapters and a brief description of each is presented in the following paragraphs.

Chapter 1 presents a background of natural organic matter in drinking water and the problems associated with it. It identifies the need for further research that is required to improve our understanding of the character of NOM and describes the main objectives of the PhD research.

A review of the characterization and influence of NOM in drinking water treatment is presented in Chapter 2. A review of the different methods that have been applied for the quantification and characterization of NOM is also presented.

In Chapter 3, results of the application of SEC-OCD and F-EEMs techniques for NOM characterization are presented. NOM in water samples from two drinking water treatment trains with distinct water quality, and from a common distribution network with no chlorine residual, was characterized and the relation between biological stability of drinking water and NOM was investigated according to concentrations of AOC.

Chapter 4 investigates the use of F-EEMs and PARAFAC to characterize NOM in drinking water treatment. The F-EEMs and SEC-OCD results presented in chapter 3 are used to investigate the relationship between the extracted PARAFAC components and the corresponding NOM SEC-OCD fractions.

Chapter 5 presents the results of NOM characterization in drinking water treatment using SEC-OCD and PARAFAC. The PARAFAC components and SEC-OCD fractions are used to evaluate the performance of two water treatment plants in terms of the removal of different NOM fractions.

In Chapter 6, the effects, on the fluorescence characteristics of NOM, of different water treatment processes for the removal of NOM in surface and ground waters are investigated. The study focuses on the fluorescence spectral shifts of a humic-like peak (peak C), at an excitation wavelength in the visible region of 300-370 nm and an emission wavelength between 400 and 500 nm, and investigates the amount of error in the determination of the fluorescence intensity maximum if the shift in the location of peak C is not taken into account.

Chapter 7 presents the results of the incorporation of fluorescence measurements, which have relatively low expense and high sensitivity and can be relatively inexpensively installed for online measurements, to improve the monitoring concentrations of DOC and total trihalomethanes (THMs) in drinking water treatment. The F-EEMs and SEC-DOC results presented in chapter 5 are employed to develop predictive models for the removal of NOM and the formation of THMs after chlorine disinfection in a full-scale drinking water treatment plant (WTP) using several water quality parameters which were measured during the period of the study.

Chapters 3 and 4 deal with results from Dutch water treatment plants which treat humic waters with high SUVA values and where no chlorine disinfection is applied. For such situations, the biological stability in the water distribution system is a critical issue. In contrast, chapters 5 and 7 deal with French water treatment plants treating waters with moderate SUVA values and applying chlorine disinfection, which could result in formation of potentially harmful DBPs.

Lastly, Chapter 8 presents a summary of the main findings and conclusions of the research study and some recommendations for practice and further research. The contents of the chapters are organized in such a manner that the results of each have been (or will be) published in international peer reviewed journals, and are generally so presented that they can be read nearly independently of the other chapters.

1.5   References

Abbt-Braun, G., Lankes, U. and Frimmel, F.H. 2004 Structural characterization of aquatic humic substances – The need for a multiple method approach. *Aquatic Sciences* 66, 151-170.

Aiken, G. and Cotsaris, E. 1995 Soil and hydrology: Their effect on NOM. *J. Am. Water Works Assoc.* 87(1), 36-45

Allpike, B.P., Heitz, A., Joll, C.A. and Kagi, R.I. 2007 A new organic carbon detector for size exclusion chromatography. *J. Chromatogr.* A 1157 472-476.

Amy, G. (1994) Using NOM Characterisation for Evaluation of Treatment.In *Proceedings of Workshop on "Natural Organic Matter in Drinking Water, Origin, Characterization and Removal"*, September 19–22, 1993, Chamonix, France. American Water Works Association Research Foundation, Denver, USA, p. 243.

Amy, G.L., Chadik, P.A. and Chowdhury, Z.K. 1987 Developing models for predcting THM formation potential and kinetics. *J. Am. Water Works Assoc.* 79, 89-97.

Chen, J., Gu, B.H., LeBoeuf, E.J., Pan, H.J. and Dai, S. 2002 Spectroscopic characterization of the structural and functional properties of natural organic matter fractions. *Chemosphere* 48(1), 59-68.

Chen, W., Westerhoff, P., Leenheer, J.A. and Booksh, K. 2003 Fluorescence Excitation-Emission Matrix Regional Integration to Quantify Spectra for Dissolved Organic Matter. *Environ. Sci. Technol.* 37, 5701-5710.

Chin, Y.-P., Aiken, G. and O'Loughlin, E. 1994 Molecular Weight, Polydispersity, and Spectroscopic Properties of Aquatic Humic Substances. *Environ. Sci. Technol.* 28, 1853-1858.

Chow, C.W.K., van Leeuwen, J.A., Drikas, M., Fabris, R., Spark, K.M. and Page, D.W. 1999 The impact of the character of natural organic matter in conventional treatment with alum. *Water Sci. Technol.* 40(9), 97-104.

Coble, P.G., Green, S.A., Blough, N.V. and Gagosian, R.B. 1990 Characterization of dissolved organic matter in the Black Sea by fluorescence spectroscopy. *Nature* 348, 432-435.

Coble, P.G., Schultz, C.A. and Mopper, K. 1993 Fluorescence contouring analysis of DOC Intercalibration Experiment samples: a comparison of techniques. *Marine Chemistry* 41, 173-178.

Croué, J.-P., G.V.Korshin and M.M.Benjamin (eds) (2000) Characterization of Natural Organic Matter in Drinking Water, AwwRF, Denver, CO.

Croué, J.-P. 2004 Isolation of humic and non-humic NOM fractions: Structural characterization. *Environmental Monitoring and Assessment* 92(1-3), 193-207.

Edzwald, J.K., Becker, W.C. and Wattier, K.L. 1985 Surrogate parameter for monitoring organic matter and THM precursors. *J. Am. Water Works Assoc.* 77, 122-132.

Fabris, R., Chow, C.W.K., Drikas, M. and Eikebrokk, B. 2008 Comparison of NOM character in selected Australian and Norwegian drinking waters. *Water Res.* 42(15), 4188–4196.

Frimmel, F.H. 1998 Characterization of natural organic matter as major constituents in aquatic systems. *Journal of Contaminant Hydrology* 35, 201–216.

Frimmel, F.H. and Abbt-Braun, G. 1999 Basic Characterization of Reference NOM from Central Europe - Similarities and Differences. *Environment International* 25(2/3), 191-207.

Her, N., Amy, G., McKnight, D., Sohna, J. and Yoon, Y. 2003 Characterization of DOM as a function of MW by fluorescence EEM and HPLC-SEC using UVA, DOC, and fluorescence detection. *Water Res.* 37, 4295–4303.

Huber, S.A., Balz, A., Abert, M. and Pronk, W. 2011 Characterisation of aquatic humic and non-humic matter with size-exclusion chromatography – organic carbon detection – organic nitrogen detection (LC-OCD-OND). *Water Res.* 45, 879-885.

Hunt, J.F. and Ohno, T. 2007 Characterization of fresh and decomposed dissolved organic matter using excitation-emission matrix fluorescence spectroscopy and multiway analysis. *J. Agricultural and Food Chemistry* 55(6), 2121-2128.

Hwang, C., Krasner, S., Sclimenti, M., Amy, G. and Dickenson, E. (eds) (2002) Polar NOM: characterization, DBPs, treatment American Water Works Association Research Foundation, Denver, CO.

Korshin, G., Chow, C.W.K., Fabris, R. and Drikas, M. 2009 Absorbance spectroscopy-based examination of effects of coagulation on the reactivity of fractions of natural organic matter with varying apparent molecular weights. *Water Res.* 43, 1541-1548.

Lee, N., Amy, G. and Croue, J.-P. 2006 Low-pressure membrane (MF/UF) fouling associated with allochthonous versus autochthonous natural organic matter. *Water Res.* 40, 2357 – 2368.

Leenheer, J.A., Croué, J.-P., Benjamin, M., Korshin, G.V., Hwang, C.J., Bruchet, A. and Aiken, G.R. (2000) *Comprehensive Isolation of Natural Organic Matter from Water for Spectral Characterizations and Reactivity Testing.* In: Natural Organic Matter and Disinfection By-Products. American Chemical Society, pp. 68-83.

Leenheer, J.A. and Croue, J.-P. 2003 Characterizing Dissolved Aquatic Organic matter: Understanding the unknown structures is key to better treatment of drinking water. *Environ. Sci. Technol.* 37(1), 19A-26A.

Leenheer, J.A. 2004 Comprehensive assessment of precursors, diagenesis, and reactivity to water treatment of dissolved and colloidal organic matter. *Water Sci. Technol. Water Supply* 4(4), 1-9.

Matilainen, A., Lindqvist, N., Korhonen, S. and Tuhkanen, T. 2002 Removal of NOM in the different stages of the water treatment process. *Environment International* 28, 457– 465.

Mopper, K. and Schultz, C.A. 1993 Fluorescence as a possible tool for studying the nature and water column distribution of DOC components. *Marine Chemistry* 41, 229-238.

Nam, S.N. and Amy, G. 2008 Differentiation of wastewater effluent organic matter (EfOM) from natural organic matter (NOM) using multiple analytical techniques. *Water Sci. Technol.* 57(7), 1009-1015.

Owen, D.M., Amy, G.L. and Chowdhary, Z.K. (eds) (1993) Characterization of Natural Organic Matter and its Relationship to Treatability, American Water Works Association Research Foundation, Denver, CO.

Owen, D.M., Amy, G.L., Chowdhury, Z.K., Paode, R., McCoy, G. and Viscosil, K. 1998 NOM characterization and treatability *J. Am. Water Works Assoc.* 87(1), 46-63.

Peuravuori, J., Koivikko, R. and Pihlaja, K. 2002 Characterization, differentiation and classification of aquatic humic matter separated with different sorbents: synchronous scanning fluorescence spectroscopy. *Water Res.* 36, 4552–4562.

Reemtsma, T., These, A., Springer, A. and Linscheid, M. 2008 Differences in the molecular composition of fulvic acid size fractions detected by size-exclusion chromatography–on line Fourier transform ion cyclotron resonance (FTICR–) mass spectrometry. *Water Res.* 42, 63-72.

Senesi, N., Miano, T.M., Provenzano, M.C. and Brunetti, G. 1989 Spectroscopic and compositional characterization of I.H.S.S. reference and standard fulvic and humic acids of various origin. *Sci. Total Environ.* 81(2), 143-156.

Stedmon, C.A., Markager, S. and Bro, R. 2003 Tracing dissolved organic matter in aquatic environments using a new approach to fluorescence spectroscopy. *Marine Chemistry* 82, 239–254.

Stedmon, C.A. and Markager, S. 2005 Resolving the variability in dissolved organic matter fluorescence in a temperate estuary and its catchment using PARAFAC analysis. *Limnol. Oceanogr.* 50(2), 686-697.

Traina, S.J., Novak, J. and Smeck, N.E. 1990 An Ultraviolet Absorbance Method of Estimating the Percent Aromatic Carbon Content of Humic Acids. *J. Environ. Qual.* 19(1), 151-153.

World Health Organisation (WHO) (2003) Managing regrowth in drinking water distribution systems. Bartram, J., Cotruvo, J., Exner, M., Fricker, C. and Glasmacher, A. (eds).

Vuorio, E., Vahala, R., Rintala, J. and Laukkanen, R. 1998 The evaluation of drinking water treatment performed with HPSEC. *Environment International* 24(5/6), 617-623.

Weishaar, J.L. 2003 Evaluation of Specific Ultraviolet Absorbance as an Indicator of the Chemical Composition and Reactivity of Dissolved Organic Carbon. *Environ. Sci. Technol.* 37, 4702-4708.

Wu, F.C., Evans, R.D. and Dillon, P.J. 2003 Separation and Characterization of NOM by High-Performance Liquid Chromatography and On-Line Three-Dimensional Excitation Emission Matrix Fluorescence Detection. *Environ. Sci. Technol.* 37, 3687-3693.

Yamashita, Y. and Jaffe, R. 2008 Characterizing the Interactions Between Metals and Dissolved Organic Matter using Excitation#Emission Matrix and Parallel Factor Analysis. *Environ. Sci. Technol.* 42, 7374-7379.

Chapter 2

CHARACTERIZATION AND INFLUENCE OF BULK NATURAL
ORGANIC MATTER (NOM) IN DRINKING WATER
TREATMENT: A REVIEW

This chapter is based on:

Baghoth, S.A., Sharma, S.K. and Amy, G. Characterization and influence of natural organic matter (NOM) in drinking water treatment: A review. In preparation for submission to *Water Research*.

2.1 Introduction

2.1.1 Background

Natural organic matter (NOM) is a heterogeneous mixture of naturally occurring compounds found abundantly in natural waters. NOM originates from living and dead plants, animals and microorganisms, and from the degradation products of these sources (Chow et al., 1999). The concentration, composition and chemistry of NOM are highly variable and depend on the sources organic matter, the physicochemical properties of the water such as temperature, ionic strength, pH and the main cation components; the surface chemistry of sediment sorbents that act as solubility control; and the presence of photolytic and microbiological degradation processes (Leenheer and Croue, 2003). NOM in general significantly influences water treatment processes such as coagulation, oxidation, adsorption, and membrane filtration and some of its constituents are particularly problematic. In addition to aesthetic problems such as color, taste and odor, it contributes to the fouling of membranes, serves as precursor for the formation of disinfection by-products, increases the exhaustion and usage rate of activation carbon and also certain fractions of NOM promote microbial growth and corrosion in the distribution system (Amy, 1994; Owen et al., 1993).

The extent to which NOM affects water treatment processes depends on its quantity and physicochemical characteristics. NOM that is rich in aromatic structures such as carboxylic and phenolic functional groups have been found to be highly reactive with chlorine, thus forming DBPs (Reckhow et al., 1990). These aromatic structures are commonly present as a significant percentage of humic substances, which typically represent over 50% of NOM. Hydrophobic and large molecular humic substances are enriched with aromatic structures and are readily removed by conventional drinking water treatment consisting of flocculation, sedimentation and filtration. In contrast, less aromatic hydrophilic NOM is more difficult to remove and is a major contributor of easily biodegradable organic carbon, which promotes microbiological regrowth in the distribution system. An understanding of the behaviour of different fractions or constituents of NOM present in water is crucial to understanding their fate and impact during water treatment and in water distribution systems.

Over the last 10–20 years, increasing NOM concentration levels in water sources have been observed in many countries due to issues such as global warming, changes in soil acidification, increased drought severity and more intensive precipitation events (Fabris et al., 2008). In addition to the trend towards increasing NOM concentration, the character of NOM can vary with source and time (season). The great seasonal variability and the trend towards elevated NOM concentration levels impose challenges to the water industry and water treatment facilities in terms of operational optimization and proper process control (Fabris et al., 2008). By systematic characterization, the problematic NOM fractions can be targeted for removal and transformation. Therefore, proper characterization of the NOM in raw water or after different treatment steps would be an important basis for selection of water treatment processes, monitoring of the performance of different treatment steps, and assessing distribution system water quality. This chapter reviews several methods that have been used to characterise bulk NOM, rather than isolates of NOM, and the influence of NOM in drinking water treatment.

2.1.2 Types and of sources NOM in drinking

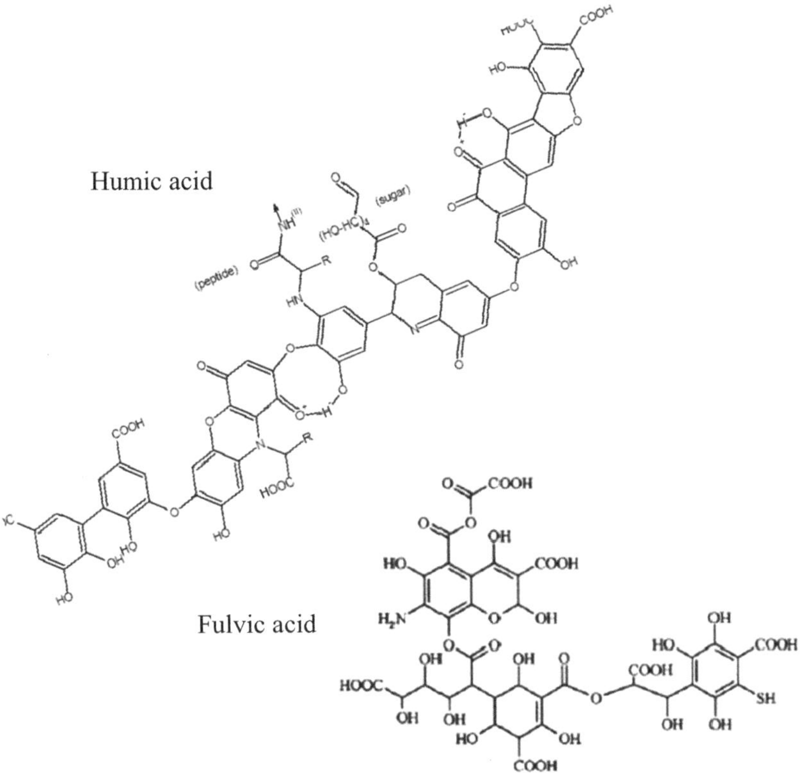

Humic acid

Fulvic acid

Figure 2.1 Proposed model molecular structure of humic and fulvic acids (Stevenson, 1982, Alvarez-Pueblaa et al., 2006).

The structural composition NOM is highly variable and depends mainly on the origin of the precursor material and the degree of modification it has undergone (Lankes et al., 2008). For example, NOM that is derived from aquatic algae has a relatively large nitrogen content and low aromatic carbon and phenolic contents, while terrestrially derived NOM has relatively low nitrogen content but large amounts of aromatic carbon and phenolic compounds (Fabris et al., 2008). Thus the aromatic fraction of NOM, which has been found to be a major reactive component, varies with different sources. DOC varies from less than 1 mg C/L in groundwater and seawater to more than 40 mg C/L in brown water and soil seepage water (Thurman, 1985). DOC concentrations in groundwater range from 0.2 to 15 mg C/L with a median concentration of 0.7 mg C/L (Thurman, 1985). Most groundwaters have concentrations of DOC below 2 mg C/L (Leenheer et al., 1974) but groundwaters recharged with organic-rich surface waters typically have higher DOC concentrations. Mean DOC concentrations in lakes depend on the trophic state of the lake and ranges from 2 mg C/L in oligotrophic, 10 mg C/L in eutrophic lakes and 30 mg C/L in dystrophic lakes (Thurman, 1985). The mean DOC concentration in rivers is 2.5 mg C/L but it varies from less than 1 mg C/L to 20 mg C/L. In drinking water treatment for removal of NOM, DOC concentrations range between 1.3 and 16 mg C/L in the source water, and between 0.8 and 5.4 mg C/L in the finished water (Allpike et al., 2005; Volk et al., 2005; Baghoth et al., 2011; Fabris et al., 2008; Hammes et al., 2010).

The NOM present in source waters used for drinking water has been classified as humic (nonpolar) and nonhumic (polar) material (Owen et al., 1993; Krasner et al., 1996). However, this operational definition of humic/nonhumic components of NOM, based on what is adsorbed (or not) on XAD resins (Malcolm and MacCarthy, 1992), has not been universally accepted. Hydrophilic ("nonhumic") fractions of NOM exhibit some of the properties typically observed for classic humic fractions (Barret et al., 2000). The operationally defined aquatic humic substances (HS) can be divided into two main fractions: humic acids (HA), which are insoluble at pH less than 1, and fulvic acids (FA), which are soluble at all pHs. Humic substances are complex macromolecules some of which consist of a mixture of many organic acids containing carboxylic and phenolic functional groups. Typical molecular structures for humic and fulvic acids are shown in Figure 2.1. Aquatic HS account for approximately 50% of the DOC present in most natural waters. The non-humic fraction of NOM consists of hydrophilic acids, proteins, amino acids, amino sugars and carbohydrates. Figure 2.2 shows a method of NOM classification that classifies DOC based on polarity (hydrophobic/hydrophilic), acid/neutral/base properties, compound class characteristics, specific compound characteristics, and compound complex characteristics (Leenheer and Croue, 2003).

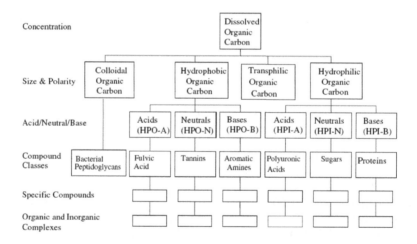

Figure 2.2 DOC fractionation diagram. (Source: Leenheer and Croue, 2003).

NOM, in general, can be divided into three main types based on the source of NOM (Sharma et al., 2011):

1) Allochthonous NOM – This type of NOM originates from the decay of terrestrial biomass or through soil leaching in the watershed, mainly from runoff or vegetative debris. The production and characteristics of this type of NOM is therefore related to vegetative patterns and to hydrologic and geological characteristics of the watershed.

2) Autochthonous NOM – This type of NOM originates from in-situ sources, mainly algal organic matter (AOM), other phytoplankton, and macrophytes; components could be excellular or intracellular organic matter consisting of macromolecules and cell fragments. The production of this type of NOM is therefore related to photosynthetic activity and decay products of algal matter.

3) Effluent organic matter (EfOM) – EfOM consists of "background" drinking water NOM which is not removed during wastewater treatment plus soluble microbial products (SMPs) formed during biological treatment of wastewater. The characteristics of EfOM therefore depend on the type of drinking water source and treatment as well as the type of wastewater treatment applied.

2.2   NOM in drinking water treatment

2.2.1 Relevance of NOM in drinking water treatment

The presence of NOM in water significantly impacts different drinking water treatment processes as well as water quality in the distribution system, leading to operational problems and increased cost of water treatment. Some of the ways in which NOM affects drinking water quality and the performance of water treatment process are summarized below:

(i)   NOM impacts aesthetic drinking water quality by imparting colour, taste and odour to the water.

(ii)   NOM increases the demand or dose of coagulants, oxidants and disinfectants required for drinking water treatment

(iii)   NOM present in water may react with chlorine or other disinfectants/oxidants to produce potentially harmful disinfection by-products (DBPs), many of which are may be carcinogenic or mutagenic.

(iv)   NOM is responsible for fouling of membranes, reducing the flux, resulting in high frequency of backwashing and cleaning of membranes to restore the flux (Jermann et al., 2007; Kimura et al., 2004).

(v)   NOM competes with target organic micropollutants for adsorption sites in activated carbon filters, adversely impacting both adsorption capacity and adsorption kinetics of the target organic micropollutants.

(vi)   Presence of biodegradable NOM in water entering the distribution system may lead to biological regrowth, when a sufficient disinfectant residual is not maintained in the distribution system (Srinivasan and Harrington, 2007; Zhang and DiGiano, 2002).

(vii)   Some NOM fractions may promote corrosion in the distribution system. Whereas some studies have shown that NOM decreases the rate of corrosion of iron pipes (Sontheimer et al., 1981; Broo et al., 1999), a study by Broo et al., 2001 found that NOM increased the rate of corrosion at low pH, but decreased it at high pH values.

2.2.2 Drinking water treatment methods for the removal of different types of NOM

The removal of NOM during drinking water treatment depends highly on the characteristics of the NOM present (e.g., molecular weight distribution (MWD), carboxylic acidity, and humic substances content), its concentration and the removal methods applied. High molecular weight (HMW) NOM is more amenable to removal than low molecular weight (LMW) NOM, particularly the fraction with an MW of 500 Dalton (Da). NOM components with the highest carboxylic functionality and hence the highest charge density are generally more difficult to remove by conventional treatment (Collins et al., 1985; Collins et al., 1986). Several water treatment methods have been used to remove NOM during drinking water treatment with varying degree of success. The following are some of the methods used:

- **Enhanced coagulation** — NOM removal in a conventional water treatment process may be achieved through the addition of a chemical coagulant. Coagulation with aluminium and iron salts is effective in the removal of NOM, as measured by total organic carbon, and removal efficiencies in the range of 25 to 70%, have been reported (Chowdhury et al., 1997; Edwards et al., 1997; Owen et al., 1996; Krasner and Amy, 1995; Owen et al., 1993; Bond et al., 2010; Abbaszadegan et al., 2007). Coagulation removes the hydrophobic fraction and high molecular weight NOM in preference to the hydrophilic fraction and low molecular weight NOM compounds (Owen et al., 1993). The former are composed of primarily humic substances (fulvic and humic acids), which are rich in aromatic carbon and phenolic structures, while the later are composed mostly of aliphatic and nitrogenous organic carbon, such as carboxylic acids, carbohydrates and proteins. Conventional water treatment involving coagulation, flocculation and sedimentation is normally optimised for removal of turbidity in raw water and to removal NOM, enhanced coagulation is required. Enhanced coagulation for the removal of NOM requires elevated coagulant doses (5-100 mg L^{-1} for Al and Fe salts), above what would be required for turbidity removal alone. Enhanced coagulation can be achieved by selection of the appropriate type of coagulant, coagulant dosage and pH for removal of a certain percentage of TOC from the raw water. However, the increased coagulant dose leads to excess sludge production and increased costs of treatment, particularly for low alkalinity waters. Enhanced coagulation is recommended for waters with hydrophobic and relatively high molecular weight NOM, as indicated by moderate to high specific ultraviolet absorbance values (SUVA). For waters with more hydrophilic and low molecular weight NOM, as well as for waters with low DOC concentrations (~ 2.0 mg CL^{-1}) and SUVA values (~ 2.0 L$(mg)^{-1}m^{-1}$), enhanced coagulation is ineffective and additional NOM removal treatment would be recommended (Volk et al., 2000).

- **Activated carbon (AC)** — Activated carbon (AC) is widely used to remove trace organic compounds from drinking water. It is an effective adsorbent for a wide range of undesirable organic compounds (e.g. pesticides and taste and odour compounds) which are often targeted for removal in drinking water treatment (Walter J.Weber, 2004). It has also been found to be effective in the removal of NOM, although NOM competes for adsorption sites with the target compounds. AC may be used as granular activated carbon

(GAC) or powdered activated carbon (PAC). GAC filters remove organic carbon through adsorption and biological degradation. In biologically active GAC filters, biodegradation is the main mechanism of organic carbon removal and the filters are made active by the absence of disinfection residual which would prevent formation of biomass that consumes the biodegradable organic carbon. In these filters, ozonation is often used prior to the GAC filters in order to degrade recalcitrant organic matter and thus promote biodegradation of the more biodegradable ozonated organic carbon. PAC is commonly applied in water treatment to remove NOM that causes odour and tastes and also to remove synthetic organic chemicals. Application of PAC reduces the levels of assimilable organic carbon (AOC) and it has been found that addition of PAC to a solids clarifier removes significantly more AOC compared to water treatment using conventional settling tanks (Camper et al., 2000). PAC is also widely used prior to ultrafiltration (UF) in order to remove NOM and thus minimize fouling of UF membranes. Adsorption of NOM by AC is controlled predominantly by the relationship between the molecular size distribution of NOM and the pore size distribution of the AC (Matilainen et al., 2006). Many studies have shown that, due to a size exclusion effect, low molecular weight (LMW) organic matter is more amenable to adsorption onto AC than high molecular weight (HMW) organic matter. When enhanced coagulation cannot sufficiently remove NOM, additional treatment by GAC filtration has been found to be effective in lowering the levels of organic carbon in the finished water.

- *Ion exchange* —Ion exchange (IEX) is an effective method for removing NOM in waters containing LMW humic substances, which are not effectively removed by coagulation. Ion exchange by electrostatic interaction is the dominant mechanism of NOM removal by IEX resins but hydrophobic interactions between the organic matter and the resin matrix can also have a significant effect on removal of specific NOM fractions. The removal of NOM by anionic exchange resins (AER) is influenced by the characteristics of the resins (strong or weak base AER), water quality (pH, ionic strength, hardness, etc.) and the character of NOM (molecular eight (MW), charge density, polarity). Since most NOM components are typically negatively charged, macroporous AER are effective for NOM removal. An innovation in ion exchange is the use of MIEX (Magnetic Ion Exchange Resin) resins which are similar to conventional resins but 2 to 5 times smaller in size (less than 180 μm). The smaller size provides a larger surface area that enhances NOM removal and improves regeneration efficiency by making it easier for NOM to diffuse in or out of the resin. However, to overcome high head loss and problems of backwashing associated with the small size of the resins, the resins are used in a continuously stirred contactor similar to a flash mixer in a conventional water treatment plant. Depending on the water quality, MIEX can remove from 30% to over 70% of the DOC in water (Humbert et al., 2005; Mergen et al., 2008; Boyer and Singer, 2005; Morran et al., 2004; Sani et al., 2008). Unlike enhanced coagulation, which removes mainly the HMW hydrophobic fraction of DOC, MIEX effectively removes the hydrophobic HMW fraction as well as the hydrophilic LMW fraction of DOC (Johnson and Singer, 2004; Allpike et al., 2005; Boyer and Singer, 2005; Mergen et al., 2009). Water treatment with MIEX has been found to remove a wider range of molecular weight and organic acids of DOC than coagulation (Morran et al., 2004; Drikas et al., 2011; Allpike et al., 2005).

- *Ozonation* — Ozonation is often used in combination with other treatment processes for NOM removal. It is often used prior to granular activated carbon (GAC) filters in order to degrade recalcitrant organic matter and thus promote biodegradation of the more biodegradable ozonated organic carbon. However, when these fractions are not well removed in biofilters or adsorbed on GAC, they tend to be more difficult to remove due to their mobility and generally increased polarity. Typically, the adsorbability of NOM decreases with ozonation because of the creation of more polar, hydrophilic compounds. The extent to which NOM is reduced in ozone enhanced biofiltration depends on several factors such as the applied ozone dose, characteristics of the NOM in the water and other water quality parameters like pH and alkalinity (Odegaard et al., 1999). Ozone preferentially reacts with the aromatic fraction of NOM, thus reducing the SUVA of the water. For NOM removal with ozone enhanced biofiltration, the ozone dose should be optimized. Specific ozone doses of 0.5 to 1.0 mg O_3/mg C are widely applied prior to biofiltation (Juhna and Melin, 2006). Increasing the ozone dose beyond 1.0 mg O_3/mg C does not significantly increase the biodegradability of NOM (Siddiqui et al., 1997). Ozonation of waters containing bromide leads to formation of bromate, a DBP and potential carcinogen, which is not removed by subsequent biofiltration.

- *Membrane filtration* — Membrane filtration systems such as ultra filtration and nanofiltration can be used to remove larger organic matter components left after coagulation and certain dissolved NOM. Ultra filtration may be used to effectively remove larger MW organic compounds but is limited by its range of molecular weight cutoff (MWCO) in effectively removing a significant fraction of lower MW organic matter. Nanofiltration membranes, which have a lower MWCO could be effectively used for removal of NOM fractions which cannot be removed by ultra filtration (Frimmel et al., 2006).

- *Bank filtration (BF)* — Bank filtration systems have been used as a pre-treatment or complete treatment of river and lake water for production of potable water. BF can remove particles, bacteria, viruses, parasites, organic compounds, and potentially nitrogen species (Kuehn and Mueller, 2000). BF is known to effectively remove bulk NOM and some organic micropollutants. BF can achieve 50% to 90% reduction of biodegradable NOM, measured as biodegradable dissolved organic carbon (BDOC) and assimilable organic carbon (AOC), and 26% reduction in SUVA values in UV absorbing NOM (Weiss et al., 2004).

- *Combined treatment processes and hybrids* — Different combination of the NOM removal methods have been employed for removal of NOM in drinking water. The main objective of these hybrid/combined systems is to maximise the removal of specific fractions of NOM more effectively. These combined treatment systems may include, (a) coagulation followed by ultra filtration, (b) ozonation followed by activated carbon filtration, (c) activated carbon filtration followed by reverse osmosis, (d) biofiltration followed by nanofiltration, (e) ion exchange followed by activated carbon filtration and (f) ozonation followed by biofiltration and membrane filtration (Owen et al., 1993; Matilainen et al., 2002; Osterhus et al., 2007; Humbert et al., 2008).

2.3 Quantification and measurement of NOM

2.3.1 Sampling and Pre-filtration

Sampling for NOM analysis should follow appropriate standard procedures (such as ASTM standards for water testing) for preparation of the sample container, sampling, sample preservation, and analysis. For reliable analysis of the samples, these four basic steps should be performed according to the procedures specified. External contamination during handling should be avoided as much as possible and for samples rich in biodegradable organic matter, rapid analysis should be carried out in order to minimise biodegradation and hydrolysis of some components of NOM. Samples that cannot be analyzed immediately after sampling should be stored at a temperature of 4°C or below.

Specially cleaned glassware should be used with hard plastic screw cups and Teflon, polypropylene (PP) or polyethylene (PE) inlays. For samples with low total organic carbon (TOC) concentrations, higher precautionary measures, such as use of glass or Teflon sampling bottles, are recommended. Other precautionary measures include (i) proper labelling of the sampling bottles, (ii) rinsing the sampling bottle with the sample, where feasible, and (iii) not freezing the samples or treating with any additives or preservatives.

Samples for dissolved organic carbon (DOC) analysis are generally filtered through 0.45 μm porosity membrane filters immediately after sampling. This filtration step also provides physical sterilization of the sample through removal of bacteria. It is necessary to use cooling boxes for the shipment of samples. The shipment method should consider the arrival time for the sample and the time between sampling and analysis. Non-cooled samples should be analysed within 24 hours after sampling (cooled: up to 72 hours). Another alternative is pasteurisation of closed sampling bottles at 70°C for 30 minutes.

2.3.2 TOC and DOC

Because of the heterogeneous character of NOM and the insignificant fraction of the TOC that trace organic contaminants in natural systems generally represent, the concentration of NOM is typically measured as the total organic concentration (TOC) in a water sample (Leenheer and Croue, 2003). Similarly, the dissolved fraction of NOM (dissolved organic matter (DOM)) is measured as dissolved organic carbon (DOC), which represents the amount of chemically reactive fraction. DOM is a complex mixture of aromatic and aliphatic hydrocarbon structures that have attached amide, carboxyl, hydroxyl, ketone, and various minor functional groups. Heterogeneous molecular aggregates in natural waters increase DOM complexity (Leenheer and Croue, 2003). Characterization of NOM typically starts with the fractionation of TOC into the operationally defined fractions of particulate organic carbon (POC), which is the fraction of the TOC retained on a 0.45 μm porosity membrane, and (DOC), which is the organic carbon smaller than 0.45 μm in diameter. DOC is chemically more reactive because it is a measure of individual organic compounds in the dissolved state, while POC is both discrete plant and animal organic matter and organic coatings on silt and clay. POC generally represents a minor fraction (below 10%) of the TOC (Thurman, 1985).

TOC and DOC concentrations are measured directly and the difference of the concentration values gives the POC concentration. DOC concentrations generally range from 0.1 milligrams carbon per litre (mg C/L) in ground water to 50 mg C/L in bogs (Thurman, 1985). Different methods for DOC/TOC analysis are available, of which the most commonly used are wet chemical oxidation and high temperature combustion (HTC). The most successful and widely used wet chemical methods are persulfate oxidation, UV irradiation, and a combination of the two (Sharp, 1993). Figure 2.3 shows the division between dissolved and particulate organic carbon, based on filtration through a 0.45 μm porosity membrane filter.

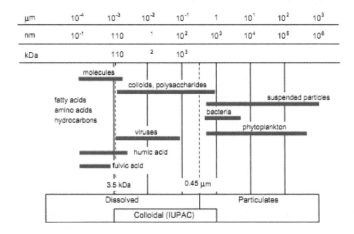

Figure 2.3 Continuum of particulate and dissolved organic carbon in natural waters (Aiken and Leenheer, 1993).

2.3.3 UVA$_{254}$ and SUVA

NOM absorbs light over a wide range of wavelengths, from ultraviolet (UV) to visible (Vis). As such, (UV/Vis) absorbance by NOM is a semi-quantitative indicator of the NOM concentration in natural waters. UV light absorbance at 254 nm (UVA$_{254}$) is widely used in water treatment plants to monitor the concentration of DOC on line, once the correlation between DOC and UVA$_{254}$ has been established for the particular water of interest (Edzwald et al., 1985; Amy et al., 1987). The effect of sample pH on the absorption of UV light was found to be minimal between pH 2.0 and 8.6 for most natural samples analyzed (Weishaar, 2003). The functional groups responsible for NOM absorbance of light are referred to as chromophores. Most of the chromophores in NOM molecules that absorb in the UV region (wavelength < 400 μm) are aromatic groups with various degrees and types of substitution, including monosubstituted and polysubstituted phenols and various aromatic acids (Traina et al., 1990; Chin et al., 1994). These UV absorbing chromophores are associated primarily with the humic fraction of NOM (Korshin et al., 1997). The reactivity of DOC and aquatic humic substances with oxidants, such as chlorine (Reckhow et al., 1990; Li et al., 2000) and ozone (Westerhoff et al., 1999), depends strongly on the aromaticity of the organic matter.

Table 2.1 Guidelines on the nature of NOM and expected DOC removals by coagulation (Source: Edzwald et al., 1985).

SUVA L/mg-m	Composition	Coagulation	DOC Removals
>4	Mostly aquatic humic High hydrophobicity High molecular weigh	NOM Controls Good DOC removal	>50% for Alum >50% for Ferric
2-4	Mixture of aquatic humics and other NOM Mixture of hydrophobic and hydrophilic NOM Mixture of molecular weights	NOM influences DOC removals OK	25-50% for Alum Little greater for Ferric
<2	Mostly non-humics Low hydrophobicity Low molecular weight	NOM has little Influence Poor DOC removal	<25% for Alum Little greater for Ferric

The most direct measurement of aromaticity of NOM is provided by ^{13}C NMR spectroscopy but it requires expensive, sophisticated instrumentation and significant sample preparation (Weishaar, 2003). SUVA provides a simpler method for estimating aromaticity of DOC in a water sample. SUVA is an "average" absorptivity for all the molecules that comprise the DOC in a water sample and has been used as a surrogate measurement for DOC aromaticity (Traina et al., 1990). Specific UV absorbance (SUVA or $SUVA_{254}$) is defined as the ratio of the samples's UV absorbance at 254 nm to the DOC concentration of the solution. There is a high correlation between SUVA and the aromatic contents of many NOM fractions (Croue et al., 1999). Waters with a high concentration of hydrophobic NOM such as humic substances have a high SUVA value and SUVA can be used to estimate the chemical characteristic of DOC in a given environment (Leenheer and Croue, 2003). Because $SUVA_{254}$ is a good indicator of the humic fraction in DOC and coagulation is effective at removing humic substances, the SUVA value of given water is an indicator of how effectively the DOC in the water can be removed by coagulation during water treatment. A SUVA $\leq$ 2 L/m-mgC indicates mostly non-humic NOM, relatively low hydrophobicity, less aromatic and of lower molecular weight, while SUVA 2-4 L/m-mgC indicates a mixture of aquatic humics, other NOM, mixture of hydrophobic and hydrophilic NOM and a mixture molecular weights (Edzwald and Tobiason, 1999). Table 2.1 shows guidelines for the interpretation of SUVA values for freshwaters and the expected removals by coagulation which were proposed by Edzwald and Tobiason (1990).

2.3.4 Differential UVA

Differential absorbance spectroscopy can be used to study the chemical characteristics of UV absorbing species (mainly humic substances) in NOM that are attached by chlorine (Korshin et al., 1999). Differential UV absorbance (ΔUVA) is defined as the change in the UV absorbance of a sample in response to any forcing function such as halogenation or ozonation (Korshin et al., 1999). Differential spectroscopy focuses on the behavior of only those chromophores that are affected by the forcing parameter. It can be used to monitor the transformation of NOM species by water treatment processes such as oxidation with chlorine or ozone. ΔUVA is measured as the difference between UV absorbance of a sample before and after a process (e.g. ozonation). Differential absorbance at wavelengths near 272nm (ΔUVA_{272}) for chlorinated surface waters has been found to strongly correlate with the concentrations of total organic halogens and those of individual disinfection by-products (Korshin et al., 1997a, 2002).

2.3.5 XAD Resin Fractionation

A DOC fractionation method was developed that classifies dissolved organics on the basis of their polarity (hydrophobic/hydrophilic), acid/neutral/base properties, compound-class characteristics, specific compound characteristics, and compound complex characteristics (Leenheer and Huffman, 1976). This method is the basis of several fractionation protocols that have been developed and used to characterize NOM in water samples by fractionating it into distinct categories using non-ionic resin sorbents such as Amberlite XAD (polymethyl methacrylate). A method that uses XAD-8 resin has been widely used to isolate humic substances, consisting of fulvic and humic acids (Thurman and Malcolm, 1981), and forms the basis for categorizing NOM in drinking water as humic or non-humic. NOM is commonly fractionated on the basis of polarity by sequential sorption chromatography on Amberlite XAD-8 and XAD-4 resins (Aiken, 1985; Leenheer et al., 2000; Aiken et al., 1992; Peuravuori and Pihlaja, 1997). By using a two column array of XAD-8 and XAD-4 resins in series, the operationally defined hydrophobic organic acids, which are less polar and composed primarily of aquatic fulvic acids, are removed from acidified water samples by sorption onto XAD-8 resins and then the operationally defined more polar hydrophilic organic acids are isolated by sorption onto XAD-4 resins. The hydrophobic organic acids may be further fractionated into fulvic acids, which are soluble at any pH, and humic acids, which precipitate at a pH lower than 2. For a diversity of aquatic NOM samples, more of the dissolved organic carbon was isolated on the XAD-8 resin (23–58%) than on the XAD-4 resin (7–25%) (Aiken et al., 1992). For these samples, the hydrophilic acids were found to have lower carbon and hydrogen contents, higher oxygen and nitrogen contents, and were lower in molecular weight than the corresponding fulvic acids. ^{13}C NMR analyses indicated that the hydrophilic acids had a lower concentration of aromatic carbon and greater hetero-aliphatic, ketone and carboxyl content than the fulvic acid.

2.3.6 Dissolved Organic Nitrogen (DON)

Dissolved organic nitrogen (DON)   comprises a relatively small fraction (0.5% to 10% by weight) of NOM in natural waters (Westerhoff and Mash, 2002) and is mainly composed of degraded amino sugars, peptides and porphyrins (Leenheer, 2004). DON is commonly the dominant form of the total dissolved nitrogen (TDN) in pristine waters but represents a lower percentage of TDN in waters impacted by human activities such as agriculture (Perakis and Hedin, 2002). DON has been found to be a major foulant of filtration membranes, exerts disinfectant/oxidant demands and may react with disinfectants to form carcinogenic nitrogenous disinfection by-products (DBPs) (e.g., haloacetonitrile (HAN), N-nitrosodimethylamine (NDMA)) (Richardson et al., 1999;Najm and Trussell, 2001; Choi and Valentine, 2002). DON also reacts with free chlorine and inorganic chloramines to form organic chloramines, which have little or no bactericidal activity (Donnermair and Blatchley Iii, 2003).

The main sources of DON in drinking water sources include upstream wastewater discharge, infiltration or runoff of organic fertilizers from agricultural areas, autochthonous biological processes such as excretion of algae products in eutrophic surface water, urban runoff, and forest litter (Tuschall and Brexonik, 1980; Alberts and Takács, 1999; Westerhoff and Mash, 2002).

DON is measured after a preparative step in which all the organic material is converted (by combustion or oxidation) to an inorganic nitrogen species (NO_2^-, NO_3^- or NH_3/NH_4^+, and/or NOx gas) prior to quantification in aqueous or gaseous phases. The dissolved material is typically operationally defined as passing a 0.45 μm filter. DON cannot be directly quantified in water due to the presence of NH_4^+, NO_2^- and/or NO_3^- present in natural waters (Westerhoff and Mash, 2002). Therefore, the determination of DON would require the measurement of the DIN concentration, the measurement of the TDN concentration and the subtraction of these two concentration values. Subtraction of several independently measured concentrations compounds analytical variances in the calculated DON concentration (Devore, 1995):

$$var([DON]) = var([TDN]) + var([NO_3^-]) + var([NO_2^-]) + var([NH_3/NH_4^+])$$

where $var([X])$ is the variance of each measurement $[X]$. In the case of high DIN/TDN ratios, the variance can be greater than the calculated DON concentration. The TDN concentration cannot be quantified directly but requires a preparatory digestion step, either oxidation or combustion, to convert all DON to DIN. A number of techniques have been used to measure DIN species, but colorimetric and ion chromatographic methods are the most common (Westerhoff and Mash, 2002). Westerhoff and Mash (2002) reviewed the available digestion methods for organic nitrogen. Three general types of DON to DIN digestion methods have been developed: wet-oxidation, high temperature oxidation and photolytic oxidation. Both wet and photolytic oxidations rely on subsequent measurement of inorganic nitrogen species in solution, while high temperature oxidation methods measure nitrogen oxide indirectly through ozonation and chemiluminescent detection. Kjeldahl digestion is widely used for analysis of wastewaters but the method cannot analyze water samples with low DON concentrations (Smart et al., 1981). Three commonly used digestion methods for the determination of TDN in water samples are the alkaline persulfate oxidation, UV oxidation and high temperature catalytic oxidation (HTCO). Results from a broad methods comparison on seawater with varying TDN and DIN concentrations showed no significant difference between the three methods (Sharp et al., 2002).

Quantifying low DON concentrations in waters with high DIN is inherently inaccurate and for waters containing elevated DIN concentrations, the indirect measurement significantly impacts the accuracy and applicability of DON measurement techniques (Westerhoff and Mash, 2002). The accuracy of DON determination can be improved by increasing the DON/TDN ratio in samples with high DIN/TDN ratios, which can be achieved by lowering the DIN concentration, or increasing the DON concentration in combination with partially removing DIN (Vandenbruwane et al., 2007). One method used to lower DIN concentration in water samples involves pre-treatment using dialysis (Lee and Westerhoff, 2005). Although ~10% of DON may be lost possibly due to the adsorption of organics onto the dialysis membrane, permeation of low molecular weight fractions or biodegradation, dialysis experiments using surface water spiked with different DIN/TDN ratios demonstrated that when DIN/TDN ratios exceeded 0.6 mg of N/mg of N, dialysis pre-treatment enabled a more accurate DON determination than when no dialysis was used Lee and Westerhoff, 2005. However, dialysis pretreatment is time consuming.

2.3.7 Fluorescence Excitation Emission Matrices (F-EEM)

2.3.7.1 Fluorescence measurements

Fluorescence spectroscopy is widely used to characterize NOM, fractions of which fluoresce when excited by UV and blue light. The fluorescence intensity and characteristics depend on the concentration and composition of NOM, as well on other factors such as pH, temperature and ionic strength of the water. Fluorescence spectroscopy permits rapid data acquisition of aqueous samples at low natural concentrations. The relatively low expense and high sensitivity of fluorescence measurements, coupled with rapid data acquisition of water samples at low natural concentrations, have made fluorescence spectroscopy using fluorescence excitation-emission matrices (F-EEM) attractive for NOM characterization of water samples. This characterization has typically involved the use of excitation-emission wavelength pairs to identify fluorophores based on the location of fluorescence peaks on F-EEM contour plots (Coble, 1996). These peaks have been used to distinguish between humic-like NOM, with longer emission wavelengths (> 350 nm), and protein-like NOM, with shorter emission wavelengths ($\leq$ 350 nm). Figure 2.4 is a typical contour plot of F-EEMs of a natural water sample showing the locations of fluorescence intensity peaks B (tyrosine-like, protein-like), T (tryptophan-like, protein-like), C (humic-like) and M (humic-like, marine humic-like) that have been previously identified (Coble, 1996). Other methods for NOM characterization using F-EEMs include: fluorescence regional integration (FRI) (Chen et al., 2003); multivariate data analysis (e.g. Principal Component Analysis, PCA, and Partial Least Squares regression, PLS) (Persson and Wedborg, 2001); and multi-way data analysis using parallel factor analysis (PARAFAC) (Stedmon et al., 2003). PARAFAC has been used to decompose F-EEMs into individual components some of which have been attributed to protein-like or humic-like NOM (Hunt and Ohno, 2007, Stedmon et al., 2003, Stedmon and Markager, 2005, Stedmon et al., 2007a, Stedmon et al., 2007b, Yamashita et al., 2008).

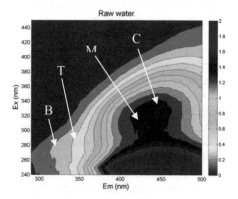

Figure 2.4 A typical surface source water fluorescence EEM contour plot showing the location of fluorescence intensity peaks B, T, M and C.

To perform fluorescence measurements, a sample is excited by a light source (such as a xenon arc lamp) and the emitted light is recorded. An excitation emission matrix (EEM) is obtained by collecting the emission spectra at a series of excitation wavelengths. The magnitude and location of the EEM peaks vary with the concentration and composition of NOM. Fluorescence intensity is known to increase with DOC, but due to the absorbance characteristic of different DOC molecules, this increase may not be linear, particularly at higher concentrations. Other light absorbing molecules or ions such as nitrate may also cause a reduction of the measured intensity. To account for these inner-filter effects, absorbance corrections have to be applied; however, these corrections are not necessary if the sample absorbance is less than 0.05 cm^{-1} or if the DOC concentration of the sample is diluted to about 1 mg C/l prior to measurement. Fluorescence is sensitive to factors such as pH, solvent polarity, temperature, redox potential of the medium, and interactions with metal ions and organic substances. To minimize metal-binding of DOC, water samples may be acidified to pH ~3 prior to measurement. However, this may lead to significant reduction in fluorescence intensities and loss of resolution for the more pH sensitive EEM peaks such as the protein-like peaks located at emission/excitation wavelengths of 320 nm/ 250 nm respectively (Westerhoff et al., 2001).

2.3.7.2   Identification of NOM fractions by PARAFAC Modelling

Parallel factor analysis (PARAFAC) may be the most useful of the available multivariate analysis techniques in investigating NOM in diverse aquatic systems (Holbrook et al., 2006). Characterization of NOM using fluorescence EEMs and parallel PARAFAC has now gained popularity as a result of recent advances that have been made in the development of algorithms for performing statistical parallel factor analysis of multilinear data. PARAFAC has been used to identify individual NOM fluorophores which have been attributed to specific fractions such as humic-like, fulvic-like and protein-like. In a study of DOM from a wide variety of aquatic environments, it was used to identify thirteen components, seven of which were attributed to quinone-like fluorophores (Cory and McKnight, 2005). These quinone-like fluorophores accounted for nearly half of the fluorescence of every sample analyzed. Quinones are common in biological molecules and are found in bacteria, some fungi and a variety of higher plant forms. They can be prepared by oxidation of aromatic amines or hydroxyls. In a different study involving characterization of NOM of surface water samples using fluorescence spectroscopy and PARAFAC, Holbrook et al., (2006) identified three different fluorophore moieties which they attributed to humic-like, fulvic-like and protein-like fractions. They also demonstrated that the characterization was consistent with expected analyte concentrations that were independently determined by wet chemistry techniques and that PARAFAC can be used to estimate (to within 30%) specific analyte concentrations in surface water. In a study to characeterize DOM in the catchment of Danish estuary using fluorescence excitation-emission spectroscopy and PARAFAC, five different fluorescent fractions were identified, four of which were attributed to allochthonous fluorescent groups and one to an autochthonous fluorescent group (Stedmon et al., 2003).

PARAFAC is a statistical model that reduces a dataset of EEMs into a set of trilinear terms and a residual array (Bro, 1997):

$$x_{ijk} = \sum_{f=1}^{F} a_{if}b_{jf}c_{kf} + \varepsilon_{ijk} \qquad i = 1,...,I; \ j = 1,...,J; \ k = 1,...,K$$

When PARAFAC is used to model EEMs, χ_{ijk} is the fluorescence intensity of the ith sample at the kth excitation and jth emission wavelength, a_{if} is directly proportional to the concentration of the fth fluorophore in the ith sample (defined as scores), and b_{jf} and c_{kf} are estimates of the emission and excitation spectrum of the fth fluorophore (defined as loadings), respectively (Stedmon et al., 2003). F is the number of fluorophore components and ε_{ijk} are the residual elements of the model. The model is fitted using an alternating least squares regression procedure. The PARAFAC model is very sensitive to the selection of number of components for fitting. Therefore, selection of the correct number of components is essential for analysing EEM data for samples of unknown fluorophore composition (Holbrook et al., 2006).

2.3.8 Size Exclusion Chromatography (SEC-DOC)

Size exclusion chromatography (SEC) is a high performance liquid chromatographic (HPLC) separation method in which the chromatographic column packing consists of precisely controlled pore sizes and the sample is fractionated according to its size or hydrodynamic volume. Larger molecules pass through without being retained and smaller molecules penetrate the pores of the packing particles and elute later. Thus, SEC chromatography can be used to fractionate NOM in a given sample according to the size of components from higher to lower, thus providing a MW or molecular size (MS) distribution. The data obtained can be represented as a SEC chromatogram in terms of either retention time or, if calibration chemicals are used, of MW distribution (Daltons). The peaks depicted in these chromatograms give an indication of the MW or MS of different components of NOM including polysaccharides (consisting of macromolecules such as polysaccharides and proteins), humic substances, building blocks (hydrolysates of humic substances), low molecular weight (LMW) acids and LMW neutrals and amphiphilic compounds.

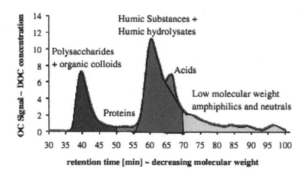

Figure 2.5 SEC-DOC chromatogram showing different fractions of NOM.

SEC has been widely used in MW/MS determinations of NOM but some results reported in the literature may be misleading because of the almost exclusive reliance on a single wavelength UV detector. While UV detection is effective for humic substances, it is less effective (or ineffective) for non-humic components of NOM such as proteins and saccharides (simple sugars and polysaccharides). Non-humic components of NOM can also be problematic in water treatment – e.g., polysaccharides are reported to be a major membrane foulant (Amy and Her, 2004; Lee et al., 2004). Other factors such as charge, molecular structure, steric effects and hydrophobicity may also influence the result (Wershaw and Aiken, 1985).

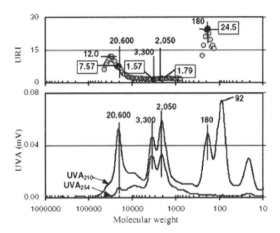

Figure 2.6 URI and SEC-UVA (at 210 nm and 254 nm) chromatograms of extracted foulant from a nanofiltration (NF) membrane (Adapted from Amy and Her, 2004).

Her et al. (2003) have coupled fluorescence and variable wavelength UV detectors (which have long been used in traditional HPLC analysis of specific organic compounds) with SEC. This permits differentiation between NOM components that exhibit high UV absorptivity at 254 nm and other NOM components that are more sensitively detected at other UV wavelengths – e.g., 210 nm for amino acids and proteins (Figure 2.5). Her et al., 2004 defined a UV absorbance (UVA) ratio index (URI) corresponding to the ratio of UVA at 210 nm to that at 254 nm (Figure 2.6). While humic substances are characterized by a URI of 1.5 to 2.0, proteins and their amino acid building blocks show higher URI values of 5 to 10.

Some NOM species fluoresce when excited by UV light and the fluorescence intensity and spectral shape depend on, besides other factors, the concentration and composition of the species. F-EEMs have been used to identify two main fluorophores (light emitting species) in NOM: humic-like and protein-like fluorophores. The pair of excitation-emission wavelengths corresponding to the fluorophore of interest can be used to collect fluorescence data by coupling a fluorescence detector sequentially with SEC chromatography. Based on 3-D F-EEM, humic substances exhibit a characteristic fluorescence intensity maximum over the excitation and emission ranges of 300-350 nm and 400-450 nm, respectively, while proteins exhibit a characteristic fluorescence intensity maximum over the excitation and emission ranges of 250-300 nm and 300-350 nm respectively (Amy and Her, 2004). Thus, depending on which NOM component is of interest, a fluorescence detector can be set to optimize recognition of either humic-like or protein-like NOM components, based on their respective EEM spectra. Fluorescence detection is, however, inappropriate for saccharides (Amy and Her, 2004).

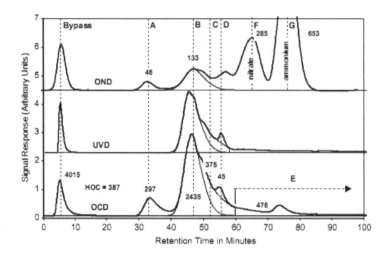

Figure 2.7 SEC-OCD chromatogram of surface water. A = Biopolymers; B = Humic substances; C = Building blocks; D = low molecular weight acids; E = low molecular weight neutrals; F, G = itrate, ammonium (only OND). HOC (hydrophobic organic carbon) = calculated difference between bypass and sum of chromatographic fractions. Values in OCD chromatogram are concentrations in mg/L C. Values in OND chromatogram are concentrations in mg/L N. (Source: Huber et al., 2011).

The use of SEC with multiple detectors, such as UV and DOC, provides a powerful analytical tool for characterizing NOM fractions from a variety of aquatic environments. Addition of an on-line DOC detector permits the detection of organic compounds which are not effectively recognized by UV detectors. SEC with multiple detectors has been shown to be very effective in following changes in the NOM distribution along drinking water treatment trains as it can capture the removal of highly reactive NOM (i.e., humic type structures) (Fabris et al., 2008), show the shift from high MW to low MW structures after oxidation processes (i.e., more biodegradable NOM) (Vuorio et al., 1998), and reveal the preferential removal of low MW DOC by biological filters (Buchanan et al., 2008). Huber and Frimmel, 1994) used a SEC system coupled with UV and a highly sensitive DOC detector (LC-OCD), based on the *Gräntzel* thin film reactor, to fractionate NOM into five fractions: biopolymers (such as polysaccharides, polypeptides, proteins and amino sugars); humic substances (fulvic and humic acids); building blocks (hydrolysates of humic substances); low molecular weight (LMW) humic substances and acids; and low molecular weight neutrals (such as alcohols, aldehydes, ketones and amino acids). The system has been further improved with the coupling of a novel organic nitrogen detector (LC-OCD-OND) which permits identification and quantification of organically bound nitrogen (e.g. bound to humic substances or biopolymers) (Huber et al., 2011). A typical LC-OCD-OND chromatogram of NOM contained in surface water is shown in Figure 2.7.

2.3.9 Biodegradable Dissolved Organic Carbon (BDOC)

Several biological methods have been developed to measure the amount of biodegradable organic matter in water. One of the most used in drinking water is the biodegradable dissolved organic carbon (BDOC) method which measures the fraction of DOC assimilated and mineralized by heterotrophic microorganisms (Huck, 1990; Servais et al., 1989; Servais et al., 1987; Frías et al., 1992; Ribas et al., 1991; Lucena et al., 1991). BDOC is based on the difference between initial and final DOC concentration after a certain period of incubation, typically 5-30 days, using an indigenous bacterial population. The methods used for BDOC determination differ mainly in the methods of inoculation, with some using suspended bacteria while others using bacteria attached to either sand or inert media in a continuous reactor column. A disadvantage of the BDOC method is the relatively long period of time it takes to obtain the results. A comparison of five methods for determination of BDOC using water samples taken across different processes of a drinking water treatment train showed that the methods did not differ significantly in their BDOC results (Frias et al., 1995). Servais et al. (1989) proposed a simple method for determination of BDOC which is applicable to surface and drinking water. It involves sterilization by filtration of the sample, inoculation with an autochthonous bacteria population and then measurement of the decrease of DOC concentration due to mineralization by bacteria. For ozonated or chlorinated water, the excess oxidant has to be neutralized prior to inoculation. The inoculated water is incubated at $20 \pm 0.5°C$ in the dark for 4 weeks. The BDOC concentration is calculated as the difference between the mean values of the initial and final DOC. Consequently, the detection limit of BDOC concentration depends on that of the DOC measurement. With current DOC analysers which have detection limits as low as 10 µg C/L, detection limits of BDOC of about 20 µg C/L are possible.

2.3.10 Assimilable Organic Carbon (AOC)

Assimilable organic carbon (AOC) is one of the most important water quality parameters with an influence on the microbiological stability of drinking water (van der Kooij, 1992). AOC is based on the measurement of biomass growth, either by plate count or by adenosine triphosphate (ATP) measurement (van der Kooij, 1992; Delahaye et al., 2003; Magic-Knezev and van der Kooij, 2004). Two strains of bacteria are commonly used in the AOC test: *Pseudomonas fluorescens* P17 and *Spirillum* NOX. However, other standard bacteria, as well as the natural flora for the samples, have been used. The AOC concentration is calculated from the maximum colony counts of the two bacteria strains, using their yield values for acetate. The AOC concentration is then expressed as acetate carbon equivalents/litre (µg C/L). A relationship between the AOC and biological stability of the water has been found and a limit of 10 µgC/L has been suggested for biostability (van der Kooij, 1992).

2.3.11 Polarity Rapid Assessment Method (PRAM)

Polarity rapid assessment method (PRAM) is a new approach developed for the characterization of NOM based on polarity using different solid-phase extraction (SPE) cartridges in parallel (Rosario-Ortiz et al., 2004). The adsorption of NOM onto each SPE sorbent is used as a measure of its total polarity and is evaluated under ambient conditions. PRAM was developed to characterize the change in NOM polarity through water treatment processes but the application was later expanded for the analysis of natural waters as well (Rosario-Ortiz et al., 2007a; Rosario-Ortiz et al., 2007b). Unlike the XAD method, which is performed at low pH and sorption is carried out in series, PRAM allows the characterization NOM polarity under ambient conditions by a series of parallel SPE sorbents of different polarity. However, the XAD method can be used to develop a mass balance, whereas the PRAM cannot. Furthermore, unlike XAD method, the PRAM cannot be used to collect NOM from samples for subsequent analysis with other techniques such as ^{13}C NMR and FTIR. Nevertheless, the use of PRAM for the characterization of NOM polarity is advantageous since it allows the analysis of NOM under ambient conditions, it requires small sample volumes (about 200 mL) and successive PRAM analyses can be completed relatively quickly, every 2 h. The different sorbents used in SPE cartridges in parallel to determine NOM adsorption include non-polar sorbents (C18, C2), which extract hydrophobic organic matter, polar sorbents (CN, silica and diol), which extract hydrophilic organic matter, and anion exchange sorbents (NH2 and SAX), which characterize the negative charge of bulk organic matter.

PRAM uses concentration (UVA_{254} or DOC) breakthrough curves normalized to the initial samples (i.e., break-through concentration values divided by the concentration of the initial sample) to provided a measure of the amount of specific NOM fractions adsorbed by the different SPE sorbents. The capacity of each SPE sorbent for specific NOM components is described as a retention coefficient (RC), which is defined as one minus the normalized maximum breakthrough level achieved Rosario-Ortiz et al., 2004. For natural and drinking water samples, UVA_{254} is used as the primary detection mechanism due to ease of use, high sensitivity, and minimum interferences. The use of DOC for quantification is hindered by excessive carbon leaching from the SPE sorbents. The levels of carbon leached were found to vary extremely within the same SPE sorbent type, as well as between different sorbents and ranged between 1–10 mg C/L (Rosario-Ortiz et al., 2004).

PRAM was used to characterize DOM from the four main tributaries of a large lake reservoir which serves as the main source of drinking water for the Las Vegas metropolitan area, the United States (Rosario-Ortiz et al., 2007b). Polarity analysis revealed clear differences in the hydrophobic/hydrophilic characters between different waters as well as temporal differences within individual waters at a particular site. It has been used to analyze the changes in NOM polarity across a pilot-scale conventional water treatment plant with pre-ozonation and biofiltration (Rosario-Ortiz et al., 2009). Changes observed in the polarity of NOM entering the pilot plant as well as across the treatment processes were attributed to changes in water blend and variability in NOM characteristics. Ozonation decreased the hydrophobicity and increased the polarity of NOM, while coagulation, flocculation and biofiltration resulted in the decrease of the hydrophobic and hydrophilic character of the chromophoric NOM. However, these polarity changes varied during the pilot plant run, and thus a more precise evaluation of NOM through unit operations should be used.

2.4   Conclusions

The operation of many drinking water treatment processes is significantly influenced by the amount and character of NOM present in water. In order to minimise the negative impacts of NOM on water treatment and to ensure that the water produced meets increasingly stringent standards, water treatment utilities need to optimize treatment processes for NOM removal. This requires a better understanding of NOM character and its removal by various treatment methods. A wide array of NOM characterization techniques have been developed which have provided considerable knowledge in understanding the impact of NOM on treatment processes. These characterization techniques differ considerably in terms of analytical approach, NOM fractionations or components analyzed, time and skills required, costs, and the form of the output or results (whether it can be interpreted easily and used by the treatment plant operators) (Chow et al., 2004).

Comparative analysis of different NOM characterization methods has demonstrated that there is no single method which can fully reveal NOM characteristics that are important for water treatment practice (Sharma et al., 2011). The use of combinations of different methods would, therefore, be required for proper analysis of the fate of different fractions of NOM during different treatment processes. In situations where high skills and costly instruments are unavailable, a basic approach of tracking DOC and SUVA changes along the treatment process train could be used to understand the removal of NOM. High performance size exclusion chromatography coupled with UV/Vis, fluorescence, light scattering and sensitive dissolved organic carbon detection techniques could be used to obtain information on molecular absorbance, size distribution, molar mass and NOM reactivity. Information on biodegradability of NOM can be obtained using bioassays to determine the concentration of BDOC or AOC.

2.5   References

Abbaszadegan, M., Mayer, B.K., Ryu, H. and Nwachuku, N. 2007 Efficacy of Removal of CCL Viruses under Enhanced Coagulation Conditions. *Environ. Sci. Technol.* 41(3), 971-977.

Aiken, G. and Leenheer, J. 1993 Isolation and Chemical Characterization of Dissolved and Colloidal Organic Matter. *Chemistry and Ecology* 8(3), 135-151.

Aiken, G.R. (1985) *Humic substances in soil, sediment, and water-Geochemistry, isolation, and characterization.* In: Isolation and concentration techniques for aquatic humic substances. Aiken, G.R., McKnight, D.M., Wershaw, R.L. and MacCarthy, P. (eds). John Wiley and Sons, New York, pp. 363-385.

Aiken, G.R., McKnight, D.M., Thorn, K.A. and Thurman, E.M. 1992 Isolation of hydrophilic organic acids from water using nonionic macroporous resins. *Organic Geochemistry* 18(4), 567-573.

Alberts, J.J. and Takács, M. 1999 Importance of humic substances for carbon and nitrogen transport into southeastern United States estuaries. *Organic Geochemistry* 30(6), 385-395.

Allpike, B.P., Heitz, A., Joll, C.A., Kagi, R.I., Abbt-Braun, G., Frimmel, F.H., Brinkmann, T., Her, N. and Amy, G. 2005 Size Exclusion Chromatography To Characterize DOC Removal in Drinking Water Treatment. *Environ. Sci. Technol.* 39(7), 2334-2342.

Alvarez-Pueblaa, R.A., Valenzuela-Calahorrob, C. and Garridoa, J.J. 2006 Theoretical study on fulvic acid structure, conformation and aggregation A molecular modelling approach. *Sci. Total Environ.* 358, 243– 254.

Amy, G. (1994) Using NOM Characterisation for Evaluation of Treatment.In *Proceedings of Workshop on "Natural Organic Matter in Drinking Water, Origin, Characterization and Removal"*, September 19–22, 1993, Chamonix, France. American Water Works Association Research Foundation, Denver, USA, p. 243.

Amy, G. and Her, N. 2004 Size exclusion chromatography (SEC) with multiple detectors: a powerful tool in treatment process selection and performance monitoring. *Water Sci. Technol. Water Supply* 4, 19-24.

Amy, G.L., Chadik, P.A. and Chowdhury, Z.K. 1987 Developing models for predcting THM formation potential and kinetics. *J. Am. Water Works Assoc.* 79, 89-97.

Baghoth, S.A., Sharma, S.K. and Amy, G.L. 2011 Tracking natural organic matter (NOM) in a drinking water treatment plant using fluorescence excitation-emission matrices and PARAFAC. *Water Res.* 45(2), 797-809.

Barret, S.E., Krasner, S.W. and AMY, G.L. (2000) *Natural organic matter and disinfection by-products characterization and control in drinking water.* American Chemical Society, Washington, D.C.

Bond, T., Goslan, E.H., Parsons, S.A. and Jefferson, B. 2010 Disinfection by-product formation of natural organic matter surrogates and treatment by coagulation, MIEX and nanofiltration. *Water Res.* 44, 1645-1653.

Boyer, T.H. and Singer, P.C. 2005 Bench-scale testing of a magnetic ion exchange resin for removal of disinfection by-product precursors. *Water Res.* 39(7), 1265-1276.

Bro, R. 1997 PARAFAC. Tutorial and applications. *Chemometrics and Intelligent Laboratory Systems* 38(2), 149-171.

Broo, A.E., Berghult, B. and Hedberg, T. 1999 Drinking water distribution-The effect of natural organic matter (NOM) on the corrosion of iron and copper. *Water Sci. Technol.* 40(9), 17-24.

Broo, A.E., Berghult, B. and Hedberg, T. 2001 Drinking water distribution-Improvements of the surface complexation model for iron corrosion. *Water Sci. Technol. Water Supply* 1(3), 11-18.

Buchanan, W., Roddick, F. and Porter, N. 2008 Removal of VUV pre-treated natural organic matter by biologically activated carbon columns. *Water Res.* (42), 3335 – 3342.

Camper, A.K., Butterfield, P., Ellis, B., Jones, W.L., Anderson, W.B., Huck, p.M., Slawson, R., Volk, C., Welch, N. and LeChevallier, M. (2000) *Investigation of the Biological Stability of Water in Treatment Plants and Distribution Systems.* American Water Works Research Foundation, Denver, CO.

Chen, W., Westerhoff, P., Leenheer, J.A. and Booksh, K. 2003 Fluorescence Excitation-Emission Matrix Regional Integration to Quantify Spectra for Dissolved Organic Matter. *Environ. Sci. Technol.* 37, 5701-5710.

Chin, Y.-P., Aiken, G. and O'Loughlin, E. 1994 Molecular Weight, Polydispersity, and Spectroscopic Properties of Aquatic Humic Substances. *Environ. Sci. Technol.* 28, 1853-1858.

Choi, J. and Valentine, R.L. 2002 Formation of N-nitrosodimethylamine (NDMA) from reaction of monochloramine: a new disinfection by-product. *Water Res.* 36, 817-824.

Chow, C.W.K., van Leeuwen, J.A., Drikas, M., Fabris, R., Spark, K.M. and Page, D.W. 1999 The impact of the character of natural organic matter in conventional treatment with alum. *Water Sci. Technol.* 40(9), 97-104.

Chow, C.W.K., Fabris, R. and Drikas, M. 2004 A rapid fractionation technique to characterise natural organic matter for the optimisation of water treatment processes. *J. Water Supply Res. Technol. AQUA* 53(2), 85-92.

Coble, P.G. 1996 Characterization of marine and terrestrial DOM in seawater using excitation-emission matrix spectroscopy. *Marine Chemistry* 51, 325-346.

Collins, M.R., Amy, G.L. and King, P.H. 1985 Removal of Organic Matter in Water Treatment. *Journal of Environmental Engineering* 111(6), 850-864.

Collins, M.R., Amy, G.L. and Steelink, C. 1986 Molecular weight distribution, carboxylic acidity, and humic substances content of aquatic organic matter: implications for removal during water treatment. *Environ. Sci. Technol.* 20(10), 1028-1032.

Cory, R.M. and McKnight, D.M. 2005 Fluorescence Spectroscopy Reveals Ubiquitous Presence of Oxidized and Reduced Quinones in Dissolved Organic Matter. *Environ. Sci. Technol.* 39, 8142-8149.

Devore, J.L. (1995) *Probability and Statistics for Engineering and the Sciences.* Brooks/Cole Publishing, Pacific Grove, CA.

Donnermair, M.M. and Blatchley Iii, E.R. 2003 Disinfection efficacy of organic chloramines. *Water Res.* 37(7), 1557-1570.

Drikas, M., Dixon, M. and Morran, J. 2011 Long term case study of MIEX pre-treatment in drinking water; understanding NOM removal. *Water Res.* 45(4), 1539-1548.

Edzwald, J.K., Becker, W.C. and Wattier, K.L. 1985 Surrogate parameter for monitoring organic matter and THM precursors. *J. Am. Water Works Assoc.* 77, 122-132.

Edzwald, J.K. and Tobiason, J.E. 1999 Enhanced coagulation: US requirements and a broader view. *Water Sci. Technol.* 40(9), 63-70.

Fabris, R., Chow, C.W.K., Drikas, M. and Eikebrokk, B. 2008 Comparison of NOM character in selected Australian and Norwegian drinking waters. *Water Res.* 42(15), 4188–4196.

Frias, J., Ribas, F. and Lucena, F. 1995 Comparison of methods for the measurement of biodegradable organic carbon and assimilable organic carbon in water. *Water Res.* 29(12), 2785-2788.

Frías, J., Ribas, F. and Lucena, F. 1992 A method for the measurement of biodegradable organic carbon in waters. *Water Res.* 26(2), 255-258.

Frimmel, F.H., Saravia, F. and Gorenflo, A. 2006 NOM removal from different raw waters by membrane filtration. *Water Sci. Technol. Water Supply* 4(4), 165-174.

Hammes, F., Berger, C., Koster, O. and Egli, T. 2010 Assessing biological stability of drinking water without disinfectant residuals in a full-scale water supply system. *J. Water Supply Res. Technol. AQUA* 59(1), 31-40.

Her, N., Amy, G., Park, H. and Von-Gunten, U. 2004 UV absorbance ratio index with size exclusion chromatography (URI-SEC) as a NOM property indicator. *J. Water Supply Res. Technol. AQUA* 57(1), 35-44.

Holbrook, R.D., Yen, J.H. and Grizzard, T.J. 2006 Characterizing natural organic material from the Occoquan Watershed (Northern Virginia, US) using fluorescence spectroscopy and PARAFAC. *Sci. Total Environ.* 361, 249–266.

Huber, S.A. and Frimmel, F.H. 1994 Direct Gel Chromatographic Characterization and Quantification of Marine Dissolved Organic Carbon Using High-Sensitivity DOC Detection. *Environ. Sci. Technol.* 28(6), 1194-1197.

Huber, S.A., Balz, A., Abert, M. and Pronk, W. 2011 Characterisation of aquatic humic and non-humic matter with size-exclusion chromatography – organic carbon detection – organic nitrogen detection (LC-OCD-OND). *Water Res.* 45, 879-885.

Huck, P.M. 1990 Measurement of biodegradable organic matter and bacterial growth in drinking water. *J. Am. Water Works Ass.* 82, 78-86.

Humbert, H., Gallard, H., Suty, H. and Croue, J.P. 2005 Performance of selected anion exchange resins for the treatment of a high DOC content surface water. *Water Res.* 39(9), 1699-1708.

Humbert, H., Gallard, H., Suty, H. and Croue, J.-P. 2008 Natural organic matter (NOM) and pesticides removal using a combination of ion exchange resin and powdered activated carbon (PAC). *Water Res.* 42, 1635 – 1643.

Hunt, J.F. and Ohno, T. 2007 Characterization of fresh and decomposed dissolved organic matter using excitation-emission matrix fluorescence spectroscopy and multiway analysis. *J. Agricultural and Food Chemistry* 55(6), 2121-2128.

Jermann, D., Pronk, W., Meylan, S. and Boller, M. 2007 Interplay of different NOM fouling mechanisms during ultrafiltration for drinking water production. *Water Res.* 41, 1713 – 1722.

Johnson, C.J. and Singer, P.C. 2004 Impact of a magnetic ion exchange resin on ozone demand and bromate formation during drinking water treatment. *Water Res.* 38(17), 3738-3750.

Juhna, T. and Melin, E. (2006) *Ozonation and biofiltration in water treatment - Operational status and optimization issues.* Techneau.

Kimura, K., Hane, Y., Watanabe, Y., Amy, G. and Ohkuma, N. 2004 Irreversible membrane fouling during ultrafiltration of surface water. *Water Res.* 38, **3431-3441**.

Korshin, G.V., Li, C.-W. and Benjamin, M.M. 1997 Monitoring the properties of natural organic matter through UV spectroscopy: A consistent theory. *Water Res.* 31(7), 1787-1795.

Korshin, G.V., Benjamin, M.M. and Li, C.W. 1999 Use of differential spectroscopy to evaluate the structure and reactivity of humics. *Water Sci. Technol.* 40(9), 9-16.

Krasner, S.W., Croué, J.-P., Buffle, J. and Perdue, E.M. 1996 Three Approaches for Characterizing NOM. *J. Am. Water Works Assoc.* 88(6), 66-79.

Kuehn, W. and Mueller, U. 2000 Riverbank filtration: an overview. *J. Am. Water Works Assoc.* 92(12), 60-69.

Lankes, U., demann, H.-D.L. and Frimmel, F.H. 2008 Search for basic relationships between ''molecular size'' and ''chemical structure'' of aquatic natural organic matter—Answers from 13C and 15N CPMAS NMR spectroscopy. *Water Res.* 42, 1051 – 1060.

Lee, W. and Westerhoff, P. 2005 Dissolved Organic Nitrogen Measurement Using Dialysis Pretreatment. *Environ. Sci. Technol.* 39(3), 879-884.

Leenheer, J.A., Malcolm, R.L., McKinley, P.W. and Eccles, L.A. 1974 Occurrence of dissolved organic carbon in selected groundwater samples in the United States. *J. Research U.S.Geol. Survey* 2, 361-369.

Leenheer, J.A. and Huffman, E.W.D., Jr 1976 Classification of organic solutes in water by using macroreticular resins. *J. Research U.S.Geol. Survey* 4, 737-751.

Leenheer, J.A. and Croue, J.-P. 2003 Characterizing Dissolved Aquatic Organic matter: Understanding the unknown structures is key to better treatment of drinking water. *Environ. Sci. Technol.* 37(1), 19A-26A.

Leenheer, J.A. 2004 Comprehensive assessment of precursors, diagenesis, and reactivity to water treatment of dissolved and colloidal organic matter. *Water Sci. Technol. Water Supply* 4(4), 1-9.

Li, C.W., Benjamin, M.M. and Korshin, G.V. 2000 Use of UV spectroscopy to characterize the reaction between NOM and free chlorine. *Environ. Sci. & Technol.* 34(12), 2570-2575.

Lucena, F., Frias, J. and Ribas, F. 1991 A new dynamic approach to the determination of biodegradable dissolved organic carbon in water. *Environmental Technology* 12(4), 343-347.

Malcolm, R.L. and MacCarthy, P. 1992 Quantitative Evaluation of XAD-8 and XAD-4 Resins used in Tandem for Removing Organic Solutes from Water. *Environment International* 18, 597-607.

Matilainen, A., Lindqvist, N., Korhonen, S. and Tuhkanen, T. 2002 Removal of NOM in the different stages of the water treatment process. *Environment International* 28, 457– 465.

Matilainen, A., Vieno, N. and Tuhkanen, T. 2006 Efficiency of the activated carbon filtration in the natural organic matter removal. *Environment International* 32, 324 – 331.

Mergen, M.R.D., Jefferson, B., Parsons, S.A. and Jarvis, P. 2008 Magnetic ion-exchange resin treatment: Impact of water type and resin use. *Water Res.* 42(8–9), 1977-1988.

Mergen, M.R.D., Adams, B.J., Vero, G.M., Price, T.A., Parsons, S.A., Jefferson, B. and Jarvis, P. 2009 Characterisation of natural organic matter (NOM) removed by magnetic ion exchange resin (MIEX Resin). *Water Sci. Technol. Water Supply* 9(2).

Morran, J.Y., Drikas, M., Cook, D. and Bursill, D.B. 2004 Comparison of MIEX® treatment and coagulation on NOM character. *Water Sci. Technol. Water Supply* 4(4), 129-137.

Najm, I. and Trussell, R.R. 2001 NDMA formation in water and wastewater. *J. Am. Water Works Ass.* 93, 92-99.

Odegaard, H., Eikebrokk, B. and Storhaug, R. 1999 Processes for the removal of humic substances from water - An overview based on Norwegian experiences. *Water Sci. Technol.* 40(9), 37-46.

Osterhus, S., Azrague, K., Leiknes, T. and Odegaard, H. 2007 Membrane filtration for particles removal after ozonation-biofiltration. *Water Sci. Technol.* 56(10), 101-108.

Owen, D.M., Amy, G.L. and Chowdhary, Z.K. (eds) (1993) Characterization of Natural Organic Matter and its Relationship to Treatability, American Water Works Association Research Foundation, Denver, CO.

Perakis, S.S. and Hedin, L.O. 2002 Nitrogen loss from unpolluted South American forests mainly via dissolved organic compounds. *Nature* 415(6870), 416-419.

Persson, T. and Wedborg, M. 2001 Multivariate evaluation of the fluorescence of aquatic organic matter. *Analytica Chimica Acta* 434(2), 179-192.

Peuravuori, J. and Pihlaja, K. 1997 Molecular size distribution and spectroscopic properties of aquatic humic substances. *Analytica Chimica Acta* 337, 133-149.

Reckhow, D.A., Singer, P.C. and Malcolm, R.L. 1990 Chlorination of Humic Materials - by-Product Formation and Chemical Interpretations. *Environ. Sci. Technol.* 24(11), 1655-1664.

Ribas, F., Frias, J. and Lucena, F. 1991 A new dynamic method for the rapid determination of the biodegradable dissolved organic carbon in drinking water. *J. Appl. Bacteriol.* 71 371-378.

Richardson, S.D., Thruston, A.D., Caughran, T.V., Chen, P.H., Collette, T.W. and Floyd, T.L. 1999 Identification of new drinking water disinfection byproducts formed in the presence of bromide. *Environ. Sci. Technol.* 33, 3378-3383.

Rosario-Ortiz, F.L., Kozawa, K., Al-Samarrai, H.N., Gerringer, F.W., Gabelich, C.J. and Suffet, I.H. 2004 Characterization of the changes in polarity of natural organic matter using solid-phase extraction: introducing the NOM polarity rapid assessment method (NOM-PRAM). *Water Sci. Technol. Water Supply* 4(4), 11-18.

Rosario-Ortiz, F.L., Snyder, S. and Suffet, I.H. 2007a Characterization of the polarity of natural organic matter under ambient conditions by the polarity rapid assessment method (PRAM). *Environ. Sci. Technol.* 41(14), 4895-4900.

Rosario-Ortiz, F.L., Snyder, S.A. and Suffet, I.H.M. 2007b Characterization of dissolved organic matter in drinking water sources impacted by multiple tributaries. *Water Res.* 41, 4115 – 4128.

Rosario-Ortiz, F.L., Gerringer, F.W. and Suffet, I.H.M. 2009 Application of a novel polarity method for the characterization of natural organic matter during water treatment. *J. Water Supply Res. Technol. AQUA* 58(3), 159-169.

Sani, B., Basile, E., Rossi, L. and Lubello, C. 2008 Effects of pre-treatment with magnetic ion exchange resins on coagulation/flocculation process *Water Sci. Technol.* 57(1).

Servais, P., Billen, G. and Hascoët, M.-C. 1987 Determination of the biodegradable fraction of dissolved organic matter in waters. *Wat. Res.* 21(4), 445-450.

Servais, P., Anzil, A. and Vantresque, C. 1989 Simple Method for Determination of Biodegradable Dissolved Organic Carbon in Water. *Appl. Environ. Microbiol.* 55(10), 2732-2734.

Sharma, S.K., Salinas Rodriguez, S.G., Baghoth, S.A., Maeng, S.K. and Amy, G. (2011) *Natural Organic Matter (NOM): Characterization Profiling as a Basis for Treatment Process Selection and Performance Monitoring.* In: Handbook on Particle Separation Processes. IWA Publishing, UK, pp. 61-88.

Sharp, J.H. 1993 The Dissolved Organic Carbon Controversy: An Update. *OCEANOGRAPHY* 6(2).

Sharp, J.H., Rinker, K.R., Savidge, K.B., Abell, J., Benaim, J.Y., Bronk, D., Burdige, D.J., Cauwet, G., Chen, W.H., Doval, M.D., Hansell, D., Hopkinson, C., Kattner, G., Kaumeyer, N., McGlathery, K.J., Merriam, J., Morley, N., Nagel, K., Ogawa, H., Pollard, C., Pujo-Pay, M., Raimbault, P., Sambrotto, R., Seitzinger, S., Spyres, G., Tirendi, F., Walsh, T.W. and Wong, C.S. 2002 A preliminary methods comparison for measurement of dissolved organic nitrogen in seawater. *Marine Chemistry* 78(4), 171-184.

Siddiqui, M.S., Amy, G.L. and Murphy, B.D. 1997 Ozone enhanced removal of natural organic matter from drinking water sources. *Water Res.* 31(12), 3098-3106.

Smart, M.M., Reid, F.A. and Jones, J.R. 1981 A comparison of a persulfate digestion and the Kjeldahl procedure for determination of total nitrogen in freshwater samples. *Water Res.* 15(7), 919-921.

Sontheimer, H., Kolle, W. and Snoeyink, V.L. 1981 The siderite model of the formation of corrosion-resistant scales. *J. Am. Water Works Assoc.* 71(11), 572-579.

Srinivasan, S. and Harrington, G.W. 2007 Biostability analysis for drinking water distribution systems. *Water Res.* 41, 2127 – 2138.

Stedmon, C.A., Markager, S. and Bro, R. 2003 Tracing dissolved organic matter in aquatic environments using a new approach to fluorescence spectroscopy. *Marine Chemistry* 82, 239–254.

Stedmon, C.A. and Markager, S. 2005 Resolving the variability in dissolved organic matter fluorescence in a temperate estuary and its catchment using PARAFAC analysis. *Limnol. Oceanogr.* 50(2), 686-697.

Stedmon, C.A., Markager, S., Tranvik, L., Kronberg, L., Slätis, T. and Martinsen, W. 2007a Photochemical production of ammonium and transformation of dissolved organic matter in the Baltic Sea. *Marine Chemistry* 104, 227–240.

Stedmon, C.A., Thomas, D.N., Granskog, M., Kaartokallio, H., Papadimitriou, S. and Kuosa, H. 2007b Characteristics of dissolved organic matter in Baltic coastal sea ice: Allochthonous or autochthonous origins? *Environ. Sci. Technol.* 41(21), 7273-7279.

Stevenson, F.J. (1982) *Humus chemistry. Genesis, composition, reactions.* John Wiley and Sons Ltd., New York, NY, USA.

Thurman, E.M. and Malcolm, R.L. 1981 Preparative isolation of aquatic humic substances. *Environ. Sci. & Technol.* 15(4), 463-466.

Thurman, E.M. (1985) *Organic geochemistry of natural waters.* Martinus Nijhoff/Dr. W. Junk Publishers, Dordrecht (Netherlands).

Traina, S.J., Novak, J. and Smeck, N.E. 1990 An Ultraviolet Absorbance Method of Estimating the Percent Aromatic Carbon Content of Humic Acids. *J. Environ. Qual.* 19(1), 151-153.

Tuschall, J.R., Jr. and Brexonik, P.L. 1980 Characterization of organic nitrogen in natural waters: Its molecular size, protein content, and interactions with heavy metals *Limnol. Oceanogr.* 25(3), 495-504.

Vandenbruwane, J., Neve, S.D., Qualls, R.G., Salomez, J. and Hofman, G. 2007 Optimization of dissolved organic nitrogen (DON) measurements in aqueous samples with high inorganic nitrogen concentrations. *Sci. Total Environ.* 386, 103–113.

Volk, C., Bell, K., Ibrahim, E., Verges, D., Amy, G. and Lechevallier, M. 2000 Impact of Enhanced and Optimized Coagulation on Removal of Organic Matter and its Biodegradable Fraction in Drinking Water. *Water Res.* 34(12), 3247-3257.

Volk, C., Kaplan, L.o.A., Robinson, J., Johnson, B., Wood, L., Zhu, H.W. and Lechevallier, M. 2005 Fluctuations of Dissolved Organic Matter in River Used for Drinking Water and Impacts on Conventional Treatment Plant Performance. *Environ. Sci. Technol.* 39, 4258-4264.

Vuorio, E., Vahala, R., Rintala, J. and Laukkanen, R. 1998 The evaluation of drinking water treatment performed with HPSEC. *Environment International* 24(5/6), 617-623.

Walter J.Weber, J. 2004 Preloading of GAC by natural organic matter in potable water treatment systems: Mechanisms, effects and design considerations. *J. Water Supply Res. Technol. AQUA* 53(7).

Weishaar, J.L. 2003 Evaluation of Specific Ultraviolet Absorbance as an Indicator of the Chemical Composition and Reactivity of Dissolved Organic Carbon. *Environ. Sci. Technol.* 37, 4702-4708.

Weiss, W.J., Bouwer, E.J., Ball, W.P., O'Melia, C.R., Aboytes, R. and Speth, T.F. 2004 Riverbank filtration: Effect of ground passage on NOMcharacter *J. Water Supply. Res. Technol. - AQUA* 53(2), 61-83.

Westerhoff, P., Aiken, G., Amy, G. and Debroux, J. 1999 Relationships between the structure of natural organic matter and its reactivity towards molecular ozone and hydroxyl radicals. *Water Res.* 33(10), 2265-2276.

Westerhoff, P., Chen, W. and Esparza, M. 2001 Organic Compounds in the Environment Fluorescence Analysis of a Standard Fulvic Acid and Tertiary Treated Wastewater. *J. Environ. Qual.* 30, 2037–2046.

Westerhoff, P. and Mash, H. 2002 Dissolved organic nitrogen in drinking water supplies: a review. *J. Water Supply Res. Technol. AQUA* 51(8).

Yamashita, Y., Jaffe´, R., Maie, N. and Tanoue, E. 2008 Assessing the dynamics of dissolved organic matter (DOM) in coastal environments by excitation emission matrix fluorescence and parallel factor analysis (EEM-PARAFAC). *Limnol. Oceanogr.* 53(5), 1900-1908.

Zhang, W. and DiGiano, F.A. 2002 Comparison of bacterial regrowth in distribution systems using free chlorine andchloramine: a statistical study of causative factors. *Water Res.* 36, 1469-1482.

Chapter 3

CHARACTERIZING NATURAL ORGANIC MATTER (NOM) IN
DRINKING WATER: FROM SOURCE TO TAP

Parts of this chapter are based on:

Baghoth, S.A., Dignum, M., Grefte, A., Kroesbergen, J. and Amy, G.L., 2010 Characterization of NOM in a
drinking water treatment process train with no disinfectant residual. *Water Sci. Technol. Water Supply* 9(4):
379-386.

Baghoth, S.A., Dignum, M., Sharma, S.K. and Amy, G.L. Characterizing natural organic matter (NOM) in
drinking water: from source to tap. Submitted to *J. Water Supply. Res. Technol. AQUA (under review)*.

Summary

Natural organic matter (NOM) in water samples from two drinking water treatment trains with distinct water qualities, and from a common distribution network with no chlorine residual, was characterized and the relation between biological stability of drinking water and NOM was investigated. NOM was characterised using fluorescence excitation–emission matrices (F-EEMs), size exclusion chromatography with organic carbon detection (SEC-OCD) and assimilable organic carbon (AOC). The treatment train with higher concentrations of humic substances produced more AOC after ozonation. NOM fractions determined by SEC-OCD, as well as AOC fractions, NOX and P17, were significantly lower for finished water of one of the treatment trains. F-EEM analysis showed a significantly lower humic-like fluorescence for that plant, but no significant differences for the tyrosine- and tryptophan-like fluorescence. For all of the SEC-OCD NOM fractions, the concentrations in the distribution system were not significantly different than in the finished waters. For the common distribution network, distribution points supplied with finished water containing higher AOC and humic substances concentrations had higher concentrations of adenosine triphosphate (ATP) and *Aeromonas* sp. The number of aeromonads in the distribution network was significantly higher than in the finished waters, whereas the total ATP level remained constant, indicating no overall bacterial growth.

3.1   Introduction

Natural organic matter (NOM) is ubiquitous in natural water sources. It is a complex mixture of heterogeneous organic compounds formed from the breakdown of plant and animal matter in the environment. Its chemical character depends on its precursor materials and the biogeochemical transformations it has undergone (Aiken and Cotsaris, 1995). It is a complex mixture of aromatic and aliphatic hydrocarbon structures with side chains of amide, carboxyl, hydroxyl, ketone and various other functional groups (Leenheer and Croue, 2003). Humic substances, which comprise of humic and fulvic acids, typically comprise 50% of NOM in most natural waters and range in molecular weight (MW) from a few hundred to 100,000 daltons (Da). NOM poses major concerns in drinking water since it causes adverse aesthetic qualities such as colour, taste and odour. It affects in a negative way the performance of water treatment processes such as granular activated carbon filtration and membrane filtration (Lee et al., 2006) and it can decrease the effectiveness of oxidants and disinfectants. Furthermore, it may produce undesirable disinfection by-products (DBPs) of health concern during oxidation processes (Owen et al., 1998). The biodegradable fraction of NOM is a carbon source for bacteria and other microorganism and may, therefore, enhance biofilm formation in water distribution networks (van der Kooij, 2003). In order to address these concerns, it is essential to limit the concentration of NOM in treated water. However, the efficiency of drinking water treatment is affected by both the amount and composition of NOM. Therefore, a better understanding of the physical and chemical properties of the various components of NOM would contribute greatly towards optimisation of the design and operation of drinking water treatment processes.

Many studies and reviews have been undertaken on the structural characterization of NOM (Chin et al., 1994; Frimmel, 1998; Baker, 2001; Abbt-Braun et al., 2004; Leenheer, 2004) but its structure and fate in drinking water treatment (individual processes and process trains) are still not fully understood. Because NOM may contain "literally thousands" of different chemical constituents, it is not realistic to characterize it on the basis of a thorough compilation of the individual compounds (Croué et al., 2000). Therefore, researchers have found it more practical to characterize NOM according to operationally defined chemical groups or fractions. Many of the characterization methods that have been used involve concentration and fractionation of NOM into groups having similar properties (Frimmel and Abbt-Braun, 1999; Peuravuori et al., 2002). However, some of these techniques have inherent inaccuracies such as may arise as a result of the overlapping of different fractions during fractionation. Furthermore, they are often laborious and time consuming and may involve extensive pre-treatment of samples which could modify the NOM character.

Other analytical techniques that can be applied to characterize bulk NOM without fractionation and with minimal sample preparation are becoming increasingly popular. Non-destructive spectroscopic techniques have been very useful in NOM characterization and offer several advantages since they require small sample volumes, are non-invasive, simple in practical application and do not require extensive sample preparation. These techniques have been previously applied for qualitative and quantitative characterization of NOM (Leenheer et al., 2000; Senesi et al., 1989). Ultraviolet (UV) absorbance and fluorescence spectrophotometric measurements are commonly used to characterize NOM. Absorption of UV and visible light (UV-Vis) by surface waters and fluorescence of natural waters are attributed to the presence within the NOM molecules of aromatic chromophores (light absorbing species) and fluorophores (fluorescent components), respectively. UV absorbance by natural water samples is correlated with the NOM concentration. As such, UV absorbance of aquatic water samples, which is typically measured at a wavelength of 254 nm (UVA_{254}), is used as a surrogate measure of the NOM concentration present. However, one drawback of UV-Vis absorbance measurements is that bulk NOM as well as NOM fractions typically exhibit nearly featureless absorption spectra, showing decreasing absorbance with increasing absorbance wavelength (Korshin et al., 2009 ; Hwang et al., 2002). The lack of peaks is attributed to overlapping absorption bands of a mixture of organic compounds in NOM and to the complex interactions between different chromophores (Chen et al., 2002). Nevertheless, UVA_{254} has been found to be a useful tool in drinking water treatment practice for on-line monitoring of dissolved organic carbon (DOC) concentration (Edzwald et al., 1985; Amy et al., 1987). Specific UV absorbance (SUVA), which is defined as the UVA_{254} of a water sample divided by the DOC concentration, and molar absorptivity at 280 nm have been found to strongly correlate with the aromaticity of a large number of NOM fractions from a variety of aquatic environments (Chin et al., 1994; Weishaar, 2003). SUVA has been used as a surrogate measure of DOC aromaticity (Traina et al., 1990) and as a surrogate parameter to monitor sites for precursors of disinfectant by-products (Croué et al., 2000).

Because it is highly sensitive, fluorescence spectroscopy is an attractive tool which has become popular for NOM characterization. A considerable amount of research has been done on NOM fluorescence of whole water samples from diverse aquatic environments (Jaffe´ et al., 2008; Coble et al., 1990; Mopper and Schultz, 1993). Fluorescence excitation-emission matrix (F-EEM) analysis, in which repeated emission scans are collected at numerous excitation wavelengths, is a fluorescence spectroscopy technique that is increasingly being used to characterize aquatic NOM (Chen et al., 2003;Wu et al., 2003; Coble et al., 1990; Coble et al., 1993; Mopper and Schultz, 1993). It has been used to characterize the natural variability in organic matter fluorescence in a groundwater based drinking water treatment plant (Stedmon et al., 2011). NOM fluorescence is commonly attributed to humic-like and protein-like fluorophores which have fluorescent signals with distinct locations of excitation and emission maxima (Mopper and Schultz, 1993; Coble, 1996). Protein-like fluorescence peaks occur at excitation and emission wavelengths similar to those of tryptophan and tyrosine amino acids. Humic-like fluorescence peaks occur at higher emission wavelengths.

Molecular weight (MW) is an important characteristic of NOM which has a strong influence on the performance of drinking water treatment processes. Low molecular weight (LMW) NOM decreases the efficiency of water treatment by activated carbon filtration as it competes for adsorption sites with target compounds. LMW NOM is generally more readily biodegradable and thus enhances the formation of biofilms in drinking water distribution systems (Hem and Efraimsen, 2001; Volk et al., 2000). MW distribution of NOM is commonly determined using high performance size exclusion chromatography (HPSEC). It is a powerful technique that has been used to characterize NOM (Her et al., 2003; Chin et al., 1994; Peuravuori and Pihlaja, 1997) and to study its fate during drinking water treatment (Chow et al., 1999; Allpike et al., 2005). Significant advancements have been made in the development of size exclusion chromatographic (SEC) separation systems and detectors for the quantification and characterization of varying apparent molecular weight (AMW) NOM fractions (Allpike et al., 2007; Nam and Amy, 2008; Reemtsma et al., 2008). The detectors used typically include on-line UV absorbance spectrophotometers. On-line DOC and fluorescence measurements of SEC NOM fractions have also been employed (Frimmel et al., 1992; Wong et al., 2002; Her et al., 2003; Allpike et al., 2007).

SEC coupled with on-line UVA_{254} and/or DOC detectors has been effectively used to follow changes in NOM distribution for water samples collected across drinking water treatment trains (Vuorio et al., 1998; Allpike et al., 2005). It has been used to show the removal of highly reactive NOM (i.e., humic structures), a shift from high to low MW structures (i.e., more biodegradable NOM) after oxidation processes, and the removal of relatively biodegradable NOM such as proteins and polysaccharides.

In this research, F-EEM and SEC with UV and DOC detectors (SEC-OCD) were used to characterize NOM in water samples from source to tap for two Dutch water treatment trains in which no chemical residual is applied in the distribution. These two complementary techniques are useful in tracking NOM fractions that are of interest in such a situation; that is, they can provide information on the fate of biodegradable NOM fractions during treatment, such as proteins, polysaccharides, and low molecular weight acids, which could influence biological stability of drinking water in the distribution network.

3.2 Methods

3.2.1 Sampling

Figure 3.1 shows the sampling points for water samples were collected from two drinking water treatment trains which supply potable water to Amsterdam and its environs. These two treatment trains, Loenderveen/Weesperkarspel (LVN/WPK) and Leiduin (LDN), are operated by Waternet Water Cycle Company. In the case of LDN, River Rhine water is pretreated by coagulation (using iron chloride) and rapid sand filtration (RSF) before being pumped to infiltration areas comprised of sand dunes. The artificial recharge of the groundwater beneath the sand dunes is accomplished by means of open infiltration channels with a total length of 25 km which ensures that there is sufficient storage for about two months of drinking water production. The infiltrated water is retained in the soil for 60 to 400 days, which further improves the quality of the water and ensures that when the water is abstracted it is of a stable quality. The water is abstracted by means of open channels which transport it to an open storage reservoir from where it is pumped to LDN water treatment plant. At LDN, the water is again treated by RSF followed by ozonation, pellet softening, biological activated carbon (BAC) filtration and slow sand filtration (SSF). The LVN/WPK water treatment train also consists of two stages: a pre-treatment plant at Loenderveen (LVN) and a post-treatment plant at Weesperkarspel (WPK), about 10 km away. At LVN, water from Bethune polder is collected, coagulated and then stored in a surface reservoir with a retention period of about 100 days. The water is then treated by RSF before being transported to WPK for post treatment, which comprises ozonation, pellet softening, (BAC) filtration and SSF.

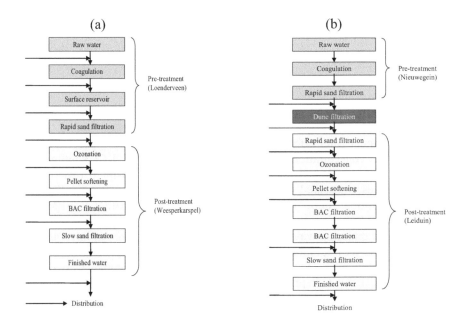

Figure 3.8 Water treatment scheme and sampling points (arrows) for the treatment process trains of (a) Loenderveen/Weesperkarspel, and (b) Leiduin.

Monthly sampling for LVN/WPK was done for the surface water source, across each treatment process, and in the distribution network. On average, twelve samples were collected from LVN/WPK process train every month in 2007. For the LDN treatment train, monthly samples were collected across the treatment processes from June to September in 2008 and in August and September 2009. For characterization of NOM in the distribution, monthly sampling was done for WPK and LDN finished waters and for three distribution points over seven months between August 2007 and February 2008. NOM was characterized using F-EEMs, which differentiates humic-like from protein-like organic matter, and SEC-OCD using a *Gräntzel* SEC-OCD system (Huber et al., 2011).

Samples were collected in duplicate and one set for analysis by SEC-OCD was transported to Het Waterlaboratorium, Haarlem, The Netherlands. The other set was transported to UNESCO-IHE Institute for Water Education for DOC, UV_{254} absorbance, SUVA and fluorescence analyses. In each case, the samples were stored under refrigeration at 5°C and the analysis performed within one week. Twelve samples were collected once a month over a

period of twelve months from January to December, 2007. Samples for SEC-OCD analysis
were collected in reusable glass bottles which were pre-cleaned by soaking in 0.01M HCl and
then in 0.1M NaOH, for 24 hours in each case. Samples for analysis at UNESCO-IHE were
collected in disposable glass bottles and pre-filtered through a 0.45 μm Whatman RC55
regenerated cellulose membrane filters within 24 hours of arrival. The pre-filtered samples
were then stored at 5 °C until required for analysis.

3.2.2 Analytical methods

3.2.2.1   DOC and UV$_{254}$ absorbance measurements

DOC concentrations of all pre-filtered samples were determined by the catalytic combustion
method using a Shimadzu TOC-V$_{CPN}$ organic carbon analyzer within one week of sampling.
UVA$_{254}$ of each sample was measured at room temperature (20±1°C) and ambient pH using a
Shimadzu UV-2501PC UV-VIS scanning spectrophotometer. SUVA was determined by
dividing the UVA$_{254}$ by the corresponding DOC concentration.

3.2.2.2   Fluorescence EEM spectroscopy

In order to minimize inner filter effects due to high DOC concentration, prefiltered samples
were diluted to DOC concentration of 1 mg C/L using 0.01 M KCl solution prior to
fluorescence measurements. The pH was then adjusted to 2.8 ± 0.1 using 0.1 M HCl and the
fluorescence intensities measured in a 1.0 cm quartz cell using a FluoroMax-3
spectrofluorometer (Horiba Jobin Yvon) at room temperature (20±1°C). EEMs for the
samples were generated by scanning over excitation wavelengths between 240 and 450 nm at
intervals of 10 nm and emission wavelengths between 290 and 500 nm at intervals of 2 nm.
An EEM of the 0.01M KCl solution was obtained and subtracted from the EEM of each
sample in order to remove most of the Raman scatter peaks. Since samples were diluted to a
DOC concentration of 1 mg C/L prior to measurements, each blank subtracted EEM was
multiplied by the respective dilution factor and then Raman-normalized by dividing by the
integrated area under the Raman scatter peak (excitation wavelength of 350 nm) of the
corresponding Milli-Q water and the fluorescence intensities reported in Raman units (RU).

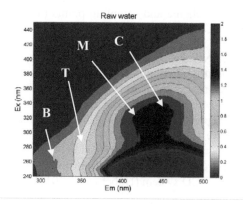

Figure 3.9 typical Loenderveen/ Weesprkarspel source water F-EEM contour plot showing
the location of fluorescence intensity peaks B, T, M and C.

In this study, fluorescence data is given as F-EEMs in which the fluorescence intensities are
given as a function of excitation and emission wavelengths. NOM characterization using F-

EEMs is done using the commonly applied approach of visual inspection of contour plots of EEMs (Coble, 1996). Table 3.1 shows the locations of fluorescence intensity peaks selected based on the F-EEMs contour plots. It also shows the corresponding peaks that were identified by Coble (1996) and that have been attributed to known fluorescing components: tyrosine-like, tryptophan-like and humic-like. The fluorescence peaks are designated by the same letters that were used previously (Coble, 1996).

Figure 3.2 is typical source water F-EEM contour plot for the LVN/WPK treatment train showing the locations of the fluorescence intensity peaks B, T, M and C. The maximum intensities of the two humic-like peaks M and C are much higher than of the protein-like peaks B and T, indicating that the source water NOM is predominantly humic in character.

Table **3.1** Selected fluorescence intensity peaks for bulk water samples.

Peak		Excitation$_{max}$ (nm)		Emission$_{max}$ (nm)		Fluorophore
Coble (1996)	This work	Coble (1996)	This work	Coble (1996)	This work	
B	B	275	280	310	320	Tyrosine-like, protein-like
T	T	275	280	340	350	Tryptophan-like, protein-like
C	C	350	330	420-480	450	Humic-like
M	M	312	310	380-420	410	Humic-like (Marine humic-like, Coble (1996))

3.2.2.3   Size exclusion chromatography with organic carbon detection (SEC-OCD)

NOM characterization using SEC-OCD was performed at Het Waterlaboratorium, Haarlem, The Netherlands. The SEC-OCD system has a detection limit of 1-50 µg C/L, which depends on the nature of the organic compound. In the system, a column TSK HW-50S is connected to a *Gräntzel* thin-film reactor (Huber and Frimmel, 1994) in which NOM is oxidized to CO_2 by UV light before it is measured by infrared detection. The column separates organic matter, according to molecular size/weight, into up to five fractions: (i) biopolymers (BP), which comprised of polysaccharides and nitrogen-containing compounds such as proteins and amino sugars, (ii) humic substances (HS), which comprised of humic and fulvic acids, (iii) building blocks (hydrolysates of humics) (BB), (iv) low molecular weight (LMW) acids and (v) LMW neutrals, such as alcohols, aldehydes, ketones and amino acids. Besides the organic carbon detector, the system also incorporates a UV detector which may be used to assess the aromaticity of the bulk NOM as well as of the humic fraction by computing the respective SUVA values. Data acquisition and processing was performed using a customised software program (ChromCALC, DOC-LABOR, Karlsruhe, Germany).

3.2.2.4   Assimilable organic carbon (AOC)

The concentration of AOC was determined by measuring the maximum level of growth of two bacterial strains, *Pseudomonas fluorescens* strain P17 and *Spirillum sp.* strain NOX, in pasteurized samples of water. After inoculation with the two bacterial strains, the water samples (600 ml) were incubated at 15°C in thoroughly cleaned 1 L glass-stoppered Erlenmeyer flasks and the growth of the bacteria in the mixed culture monitored using periodic colony counts. The AOC concentration, expressed in μg of acetate-C equivalent per litre, was calculated from the maximum colony counts attained in the samples (van der Kooij, 1992).

3.2.2.5 Aeromonas 30°C

Aeromonas sp. from water samples were counted after 42 h growth at 30°C on Ampicillin-dextrin agar (Havelaar et al., 1987).

3.2.2.6 Adenosine triphosphate (ATP)

Concentrations of adenosine triphosphate (ATP) in water samples were measured according to standard procedures (Delahaye et al., 2003; Magic-Knezev and van der Kooij, 2004).

3.3 Results and discussion

3.3.2 Bulk water characteristics for LVN/ WPK water treatment train.

Table 3.2 DOC concentration, pH, UVA_{254} and SUVA across Loenderveen/ Weesperkarspel drinking water treatment train.

Sample	pH	DOC^a (mg C/L)	$UVA_{254}{}^a$ (1/m)	$SUVA^a$ (L/mg/m)
Raw water	7.9 ± 0.3	9.0 ± 0.8	27.1 ± 3.3	3.5 ± 0.3
Coagulation effluent	7.7 ± 0.3	7.1 ± 0.6	19.1 ± 1.6	3.0 ± 0.1
Surface reservoir effluent	7.8 ± 0.3	6.5 ± 0.2	16.1 ± 0.5	3.0 ± 0.1
RS filtration effluent	7.8 ± 0.3	6.0 ± 0.3	15.1 ± 0.7	2.8 ± 0.1
Ozonation effluent	7.8 ± 0.3	5.7 ± 0.3	9.0 ± 0.6	1.8 ± 0.1
Pellet softening effluent	7.9 ± 0.3	5.4 ± 0.3	8.7 ± 0.9	1.7 ± 0.1
BAC filtration effluent	8.1 ± 0.2	3.0 ± 0.5	3.9 ± 0.9	1.5 ± 0.1
Finished water	8.1 ± 0.2	2.7 ± 0.3	3.8 ± 0.6	1.5 ± 0.1

[a] Mean value ± standard deviation, for $n = 13$.

Table 3.2 shows a summary of the means and standard deviations of the pH, DOC concentration, UVA_{254} and SUVA of water samples collected from LVN/WPK drinking water treatment train. The DOC concentration of source water from the pre-treatment plant at LVN varied from a minimum of 7.6 mg C/L in autumn, to a maximum of 9.8 mg C/L in winter, with a monthly average of 9.0 mg C/L. Coagulation with ferric chloride reduced the DOC concentration to 7.1 mg C/L. Retention in a surface reservoir for a period of about 100 days followed by RSF, the final treatment step during pre-treatment at LVN, reduced the DOC concentration to 6.0 mg C/L. Thus, pre-treatment reduced DOC concentration by about 33%. At the post-treatment plant of WPK, the DOC concentration was reduced to 2.7 mg C/L, corresponding to a removal efficiency of about 56%. However, whereas the mean percentage removal of DOC during pre-treatment was nearly half as much as during post-treatment, the actual mean DOC removal was nearly the same: 3.0 mg C/L during pre-treatment and 3.3 mg C/L during post-treatment. The average SUVA for LVN/ WPK source water was 3.5 L/mg-m, which is representative of NOM of moderate aromaticity, while that of the finished water was 1.5 L/mg-m, which is typical of NOM with low aromaticity (SUVA<2 L/mg-m). Thus, the LVN/WPK treatment train significantly reduced the aromatic character of NOM in the treated water.

Figure 3.3 shows the temporal variation of DOC concentration of the raw water as well as of effluent from the different processes of the LVN/WPK water treatment train. The seasonal variations were generally slight. The variation was more pronounced for the raw water and the pre-treatment by coagulation, showing generally higher levels during summer than during winter. However, this seasonality was effectively dampened after about three months of storage in the surface reservoir. During post-treatment, it is mainly BAC filtered and finished waters that showed a similar seasonal variation but with slightly less DOC in summer than in winter, an indication of the better performance of BAC filtration during warmer periods.

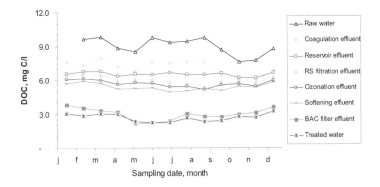

Figure 3.10 Annual variation of DOC concentration of raw water and across the Loenderveen/Weesperkasrpel treatment train.

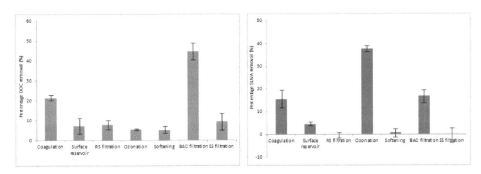

Figure 3.11 Incremental percentage removal of DOC (left) and reduction of SUVA (right) across the water treatment processes of the Loenderveen/Weesperkarspel water treatment train. Error bars indicate the 95% confidence interval of the mean (12 degrees of freedom).

Figure 3.4 shows the mean incremental percentage removal of DOC and reduction of SUVA across the water treatment processes of the LVN/WPK treatment train. DOC is mainly removed by two treatment processes: coagulation during pre-treatment, which removes, on average, about 21% of the influent DOC; and BAC filtration during post-treatment, which removes, on average, about 45% of the BAC filter influent DOC. SUVA is reduced mainly by coagulation in the pre-treatment at LVN, and by ozonation and BAC filtration in the post-treatment at WPK. Coagulation achieves nearly as much percentage reduction of SUVA

(16%) as of DOC (21%), indicating that mostly humic substances, which are more amenable
to removal by coagulation, are removed. Other studies have shown that larger molecular
weight hydrophobic humic substances are preferentially removed by coagulation (Bolto et al.,
2002; Volk et al., 2000). The percentage removal of DOC by BAC filtration (45%) is more
than two and a half times that of SUVA (17%), indicating that a significant fraction of the
DOC removed by BAC filtration is of lower aromaticity. In contrast, the percentage reduction
of SUVA by ozonation (38%) is more than six times that of DOC (6%). This can be ascribed
to the transformation of larger and more aromatic humic substances to smaller and less
aromatic humic substances, BBs and LMW acids, rather than to intact removal of DOC by
ozonation.

3.3.3   Characterizing NOM using SEC-OCD−LVN/ WPK water treatment train.

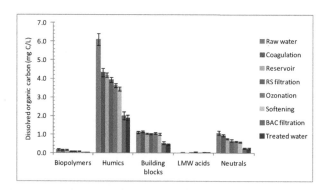

Figure 3.12 DOC concentration of SEC-OCD fractions for water samples collected across
each treatment process of Loenderveen/Weesperkarspel water treatment train. Error bars
indicate the 95% confidence interval of the mean (12 degrees of freedom).

Figure 3.5 shows the mean DOC concentration of each SEC-OCD fraction for samples
collected across the LVN/WPK treatment train. For all of the water samples analysed, humic
substances were dominant. Humics substances ranged from a maximum of 6.1 mg C/L to a
minimum of 1.9 mg C/L in the raw and finished waters, respectively. Building blocks ranged
from a maximum of 1.1 mg C/L in the raw water to a minimum of 0.5 mg C/L in the finished
water. Neutral compounds ranged from a maximum of 1.1 mg C/L in the raw water to a
minimum of 0.2 mg C/L in the finished water. Biopolymers and low molecular acids were
generally present at very low concentrations, with the biopolymers ranging from a maximum
of 0.2 mg C/L in the raw water to less than 0.1 mg C/L in the finished water, while LMW
acids were only detected after ozonation and at concentrations of less than 0.1 mg C/L. In
terms of percentage fractional composition to the NOM pool, there was no statistically
significant difference in the humic substances contribution of the DOC in the raw and
finished waters. The percentage contribution was 72% in raw and 73% in finished waters,
respectively. However, the contribution dropped significantly to 66% after coagulation,
which preferentially removes larger MW humic substances, but then increased to 70% after
BAC filtration, which preferentially removes LMW non-humic organics. While the mean
percentage contribution of both the biopolymers and the neutral compounds decreased after
treatment, that of building blocks increased as a result of the combined effect of the
preferential removal of the humic fraction by coagulation and the slight increase of the

building blocks after ozonation. The LMW acids were not measureable in the raw water but were detected after ozonation but at concentrations of less than 0.1 mg C/L. The mean percentage contribution of the LMW acids was about 1% after ozonation and then decreased to less than 1% in the finished water.

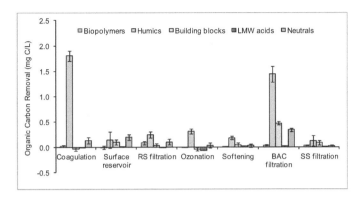

Figure 3.13 Removal of SEC-OCD fractions across Loenderveen/Weesperkarspel water treatment train. Error bars indicate the 95% confidence interval of the mean (12 degrees of freedom).

Figure 3.6 shows the removal of the five SEC-OCD NOM fractions by the treatment processes across LVN/WPK water treatment train. The humic fraction, which contributed 72% of the total DOC in the raw water, was removed mainly by coagulation and BAC filtration. Coagulation and BAC filtration removed 1.8 mg C/L and 1.4 mg C/L, respectively, representing 30% and 42% of the influent humic fraction DOC, respectively. Thus, after the breakdown of larger MW humic substances by ozonation, a significant fraction of the humic fraction was readily biodegradable and was effectively removed by BAC filtration. Since the RSF treats water previously pre-treated by conventional treatment and sand dune infiltration, and a long term analysis of the DOC removal by RSF shows a strong seasonal fluctuation (data not shown), the DOC removal by RSF is mainly biological. Building blocks were removed mostly by BAC filtration, which removed 47% of the influent building blocks fraction DOC. Biopolymers were significantly removed by all the three filtration processes, RSF (44%), BAC filtration (46%) and SSF (69%). LMW acids, which were formed after ozonation, were largely removed by pellet softening, BAC filtration and SSF. Neutral compounds were removed mostly by BAC filtration (57%) and, to a lesser extent, by coagulation (11%), surface water storage (19%) and RSF (14%).

3.3.4   Characterizing NOM using F-EEMs − LVN/ WPK water treatment train.

Figure 3.7 shows F-EEM contour plots for raw, ozonated and finished water samples collected from LVN/WPK water treatment train on 8th May 2007. The fluorescence characteristics of all of the samples were characterized by a broad humic-like peak that is typical of all natural waters. NOM characterization using F-EEMs is based on the average intensities of the fluorescence peaks identified in Section 3.2.2.2. Humic-like peak C is characteristic of most natural waters (Coble, 1996, Baker, 2001) and is characterized, in this study, by an excitation maximum at 330 nm and an emission maximum at 450 nm. Humic-like peak M has an excitation maximum at 310 nm and an emission maximum at 410 nm. This fluorescence was initially thought to originate only from marine environments but has now been found to be present in freshwaters influenced by agricultural inputs (Stedmon and Markager, 2005). Protein-like peak T has fluorescence characteristics similar to that of amino acid tryptophan (Coble, 1996) and has been attributed to fluorescence of tryptophan present in protein structures (Yamashita and Tanoue, 2003). In this study, peak T is characterized by an excitation maximum at 280 nm and an emission maximum at 350 nm. Protein-like peak B has spectral characteristics similar to that of amino acid tyrosine and is characterized in this study by an excitation maximum at 280 nm and an emission maximum at 320 nm.

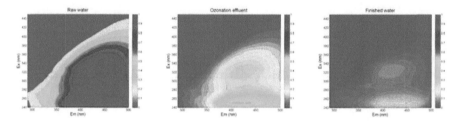

Figure 3.14 F-EEM contour plots for raw, ozonated and finished waters for Loenderveen/ Weesperkarsel water treatment train (sampling of 08/05/2007).

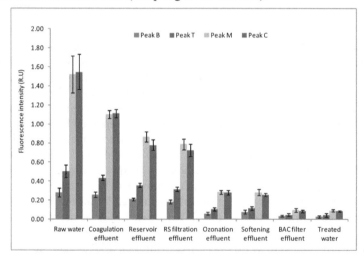

Figure 3.15 Variation of fluorescence intensities of peaks B, T, M and C across Loenderveen/ Weesperkarsel water treatment train. Error bars indicate the 95% confidence (12 degrees of freedom).

Figure 3.8 shows the variation of the mean fluorescence intensities of the four fluorescence peaks across the LNV/WPK treatment train. Consistent with the results of SEC-OCD, which showed that, on average, humic substances comprised more than 65% of DOC for all samples analyzed, the two humic-like peaks C and M were dominant for all samples across the treatment train. For protein-like fluorescence, the mean fluorescence intensity of tryptophan-like peak T was significantly ($p < .05$) higher than that of tyrosine-like peak B for all of the samples across the treatment train. For humic-like fluorescence, the mean fluorescence intensity of peak M was significantly ($p < .05$) higher than of peak C for all of the samples except for raw and coagulated waters. For raw and coagulated waters, there was no significant difference between the mean fluorescence intensity of peaks M and C.

Figure 3.9 shows the mean incremental percentage reduction in the fluorescence intensities of the protein-like peaks B and T, and of the humic-like peaks M and C. Fluorescence intensities may be reduced in two ways: (i) intact removal of fluorescent components by, for example, coagulation and BAC filtration or (ii) transformation of fluorescent components by, for example, ozonation. In this study, the changes in NOM fluorescence properties occurred mainly during ozonation and BAC filtration. Ozonation decreased the fluorescence intensities of protein-like peaks B and T by 69% and 67%, respectively. The fluorescence intensities of humic-like peaks M and C were reduced by 64% and 61%, respectively. The high percentage reduction in fluorescence intensities of all of the four peaks by ozonation could be explained by the fact that proteins as well as humic substances consist of unsaturated bonds which could be broken through oxidation by ozonation. BAC filtration decreased protein-like peaks B and T by 56% and 58%, respectively, and humic-like peaks M and C by 66% and 68%, respectively. Both ozonation and BAC reduced fluorescence intensities of all of the four peaks in preference to DOC concentration, which was reduced by 6% and 45% by ozonation and BAC respectively. The preferential reduction of fluorescence by ozonation is explained by the fact that, unlike BAC, which removes DOC intact, by adsorption, or by mineralization through biodegradation, ozonation only slightly mineralizes DOC and mainly transforms large MW NOM into smaller and less aromatic organic compounds (Swietlik et al., 2004).

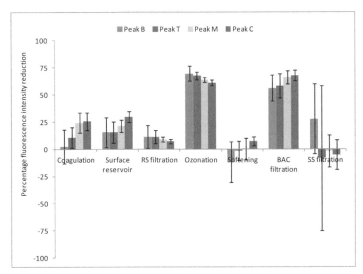

Figure 3.16 Incremental percentage reduction of fluorescence intensity of peaks B, T, M and C across Loenderveen/Weesperkarspel water treatment train. Error bars indicate the 95% confidence interval of the mean (12 degrees of freedom).

3.3.5 Bulk water characteristics for LDN water treatment train.

Table 3.3 shows a summary of the means and standard deviations of the pH, DOC concentration, UVA_{254} and SUVA of water samples collected from the LDN drinking water treatment train. The mean DOC concentration of the source water, previously pre-treated by coagulation, RSF and infiltration in sand dunes, was 2.5 mg C/L. After treatment by RSF, ozonation, pellet softening, BAC filtration and SSF, the final DOC concentration of the finished water was 1.0 mg C/L. Thus, the LDN water treatment reduced DOC concentration by 57%, which is similar to the reduction by the WPK post-treatment plant (56%). The mean SUVA for LDN pre-treated water was 2.7 L/mg/m, while that of the finished water was 1.2

L/mg/m. Thus, the LDN treatment train also significantly reduced the aromatic character of NOM in the treated water.

Table 3.3 DOC concentration, pH, UVA$_{254}$ and SUVA across Leiduin drinking water treatment train.

Sample	pH	DOC[a] (mg C/L)	UVA$_{254}$[a] (1/m)	SUVA[a] (L/mg/m)
Pretreated water	8.0 ± 0.1	2.5 ± 0.4	6.9 ± 0.9	2.7 ± 0.4
RS filtration effluent	7.9 ± 0.2	2.1 ± 0.3	5.5 ± 0.7	2.6 ± 0.2
Ozonation effluent	8.0 ± 0.1	2.0 ± 0.2	3.6 ± 0.3	1.8 ± 0.2
Pellet softening effluent	7.9 ± 0.1	1.9 ± 0.3	3.4 ± 0.2	1.7 ± 0.2
BAC filtration effluent	8.1 ± 0.3	1.2 ± 0.2	1.5 ± 0.3	1.2 ± 0.1
Finished water	8.2 ± 0.2	1.0 ± 0.1	1.3 ± 0.2	1.2 ± 0.2

[a] Mean value ± standard deviation, for $n = 7$.

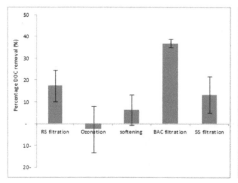 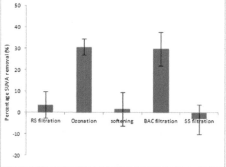

Figure 3.17 Incremental percentage removal of DOC (left) and reduction of SUVA (right) across the water treatment processes of Leiduin water treatment train. Error bars indicate the 95% confidence interval of the mean (6 degrees of freedom).

Figure 3.10 shows the mean incremental percentage removal of DOC and reduction of SUVA across the LDN water treatment processes. DOC is mainly removed by the three filtration processes, RSF (17%), BAC filtration (37%) and SSF (13%). SUVA is reduced mainly by ozonation (31%) and by BAC filtration (30%).

3.3.6 Characterizing NOM using SEC-OCD−LDN water treatment train.

Figure 3.11 shows the mean DOC concentration of each SEC-OCD fraction for samples collected across the LDN water treatment train. For all of the water samples analysed, humic substances were dominant. The concentration of humic substances ranged from a maximum of 1.5 mg C/L in the pre-treated water to a minimum of 0.6 mg C/L in the finished water. BBs ranged from a maximum of 0.4 mg C/L in the raw water to a minimum of 0.2 mg C/L in the finished water. Neutral compounds ranged from a maximum of 0.4 mg C/L in the raw water to a minimum of 0.1 mg C/L in the finished water. BPs and LMW acids were generally present at very low concentrations, with the BPs ranging from a maximum of 0.2 mg C/L in the raw water to less than 0.1 mg C/L in the finished water, while the LMW acids were only detected after ozonation and at concentrations of less than 0.1 mg C/L. The percentage fractional contribution of humic substances to the NOM pool increased slightly as a result of treatment. It increased from 58% in the pre-treated to 60% in the finished water, which is significantly less than for the LVN/WPK water treatment train, which was about 70% for both raw and finished waters. While the mean percentage contribution of both the BPs and the neutral compounds decreased after treatment, that of BBs increased, partly as a result of the slight increase of the BBs after ozonation. The concentration of LMW acids was below the detection limit of the SEC-OCD system for all of the samples.

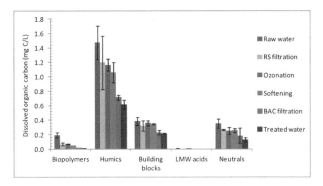

Figure 3.18 Dissolved organic carbon concentration of SEC-OCD fractions for water samples collected across the Leiduin water treatment processes. Error bars indicate the 95% confidence interval of the mean (6 degrees of freedom).

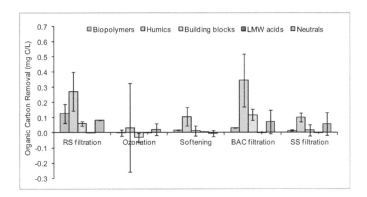

Figure 3.19 Removal of SEC-OCD fractions across Leiduin water treatment train. Error bars indicate the 95% confidence interval of the mean (6 degrees of freedom).

Figure 3.12 shows the removal of the five SEC-OCD NOM fractions by the treatment processes across the LDN water treatment train. The humic fraction, which contributed 58% of the total DOC in the pre-treated water, was removed mainly by RSF and BAC filtration. RSF removed 0.27 mg C/L, while BAC filtration removed 0.35 mg C/L, corresponding to 19% and 32% removal efficiency of humic substances, respectively. Thus, after the breakdown of larger molecular weight humic substances by ozonation, a significant fraction of the humic fraction was readily biodegradable and was effectively removed by BAC filtration. BBs were removed mostly by BAC filtration, which removed 34% of the influent BBs DOC. BPs were removed mostly by RSF upstream of the treatment train, which removed 65% of the influent BPs DOC fraction. Neutral compounds were nearly equally

removed by the three filtration processes, with RSF removing 23%, BAC filtration 29% and
SSF 27% of the influent concentration.

3.3.7   Characterizing NOM using F-EEMs—LDN water treatment train.

Figure 3.13 shows F-EEM contour plots for pre-treated, ozonated and finished water samples
collected from the LDN water treatment train on 25th August 2009. As in the case for the
LVN/WPK water treatment train, the fluorescence characteristics of all of the samples were
characterized by a broad humic-like fluorescence peak that is typical of all natural waters.

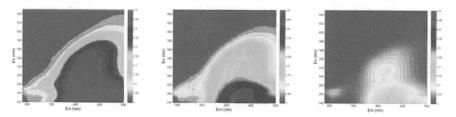

Figure 3.20 F-EEM contour plots for pre-treated, ozonated and finished waters for Leiduin
water treatment train (sampling of 25/08/2009).

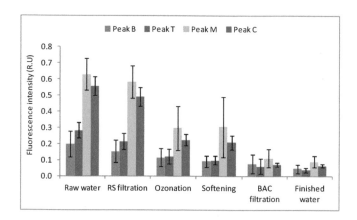

Figure 3.21 Variation of fluorescence intensities of peaks B, T, M and C across Leiduin
water treatment train. Error bars indicate the 95% confidence interval of the mean (6 degrees
of freedom).

Figure 3.14 shows the variation of the mean fluorescence intensities of the four fluorescence
peaks across the LDN treatment train. Consistent with the results of SEC-OCD, which
showed that, on average, humic substances comprised more than 58% of DOC for all samples
analyzed, the two humic-like peaks C and M were dominant for all samples across the
treatment train. For protein-like fluorescence, the mean fluorescence intensity of tryptophan-
like peak T was not significantly different from that of tyrosine-like peak B for all of the
samples across the treatment train except for raw and RSF samples. For both raw and RSF
samples, the mean fluorescence intensity of peak T was slightly higher than that of peak B.
For humic-like fluorescence, the mean fluorescence intensity of peak M was higher than that
of peak C for raw, RSF and treated water samples. However, the difference between the mean

fluorescence intensities for peaks M and C were not significant ($p = .05$) for ozonation, softenning and BAC filtration samples.

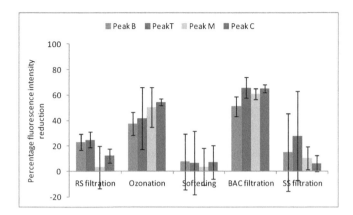

Figure 3.22 Incremental percentage reduction of fluorescence intensity of peaks B, T, M and C across Leiduin water treatment train. Error bars indicate the 95% confidence interval of the mean (6 degrees of freedom).

Figure 3.15 shows the mean percentage reduction in the fluorescence intensities of the protein-like peaks B and T, and the humic-like peaks M and C. As was the case for the LVN/WPK water treatment train, the changes in NOM fluorescence properties occurred mainly during ozonation and BAC filtration. However, RSF did not significantly reduce the fluorescence intensity of the protein-like peaks B and C in the case of LVN/WPK but increased it by at least 20% in the case of the LDN treatment train. For LVN, ozonation decreased the fluorescence intensities of the protein-like peaks B and T by 37% and 53%, respectively. The fluorescence intensities of the humic-like peaks M and C were reduced by 47% and 54%, respectively. The high percentage reduction in fluorescence intensities of all of the four peaks by ozonation could be explained by the fact that proteins as well as humic substances consist of unsaturated bonds which could be broken through oxidation by ozonation. BAC filtration decreased protein-like peaks B and T by 51% and 66%, respectively, and humic-like peaks M and C by 62% and 65%, respectively. However, for both ozonation and BAC filtration, the mean percentage reduction of fluorescence intensities of protein-like peak T and humic-like peaks M and C were not significantly ($p > .05$) different.

3.3.8   Characterizing NOM in the distribution network

The drinking water distribution network of Amsterdam city and its environs is served by the two drinking water treatment trains, LVN/WPK and LDN. A number of different water quality parameters are routinely measured for finished waters of the two treatment trains and the results consistently demonstrate that there is a significant difference between the two treatment trains in all the water quality parameters likely to influence biological stability of the water in the distribution network. The concentration NOM in the source water, measured as DOC, is generally higher for the LVN/WPK treatment train than for LDN, thus resulting in a higher DOC concentration of the finished water for the former. Furthermore, ozonation of water with higher DOC results in a higher concentration of AOC. Results from routine measurements indicate that DOC and AOC concentrations of finished waters for LVN/WPK are typically more than twice as high as for LDN. Generally, AOC concentration of the finished water for LDN is less than 10 µg ac-C.L^{-1}, the threshold for biologically stable water (van der Kooij, 1992), but more than 10 µg ac-C.L^{-1} for LVN/WPK. Mean values of some of the routinely measured water quality parameters for the year 2007 are shown in Table 3.4.

Table 3.4 Mean values of water quality parameters of finished waters for Weesperkarspel (WPK) and Leiduin (LDN) water treatment plants for the year 2007.

Parameter	Units	LDN	WPK
pH		8.17 ± 0.02	7.92 ± 0.03
TOC	(mg C/L)	1.19 ± 0.08	2.89 ± 0.09
AOC	(µg ac-C/L)	6.4 ± 1.1	18.4 ± 3.2
Aeromonas spp. of finished water	(cfu.l^{-1})	19 ± 8	390 ± 110
Aeromonas spp. in distribution system	(cfu.l^{-1})	730 ± 280	8220 ± 1270

Since historical data had demonstrated significant differences between LVN/WPK and LDN treatment trains in terms of water quality parameters likely to influence biological stability of drinking water in the distribution system, the relation between NOM characteristics and biological stability of water in the distribution network was investigated over a period of seven months. Finished water from the two treatment trains as well as water samples from three points in the distribution were collected monthly. The following distribution network sampling points were selected: Distribution 1 (DN1), which is along the distribution mains from Weesperkarspel (WPK) treatment plant; Distribution 2 (DN2), is in a zone which is supplied by both treatment trains, with the composition dependant on the hydraulic conditions within the distribution network; Distribution 3 (DN3), which is in a zone supplied by Leiduin treatment plant. Particle counters and a hydraulic model of the distribution system gave indications of residence times for the three distribution points as 15-16 hours for DN1, 22-35 hours for DN2 and 33 hours for DN3.

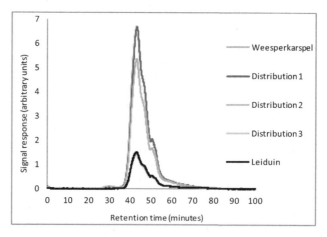

Figure 3.23 SEC-OCD chromatogram of finished waters of Weesperkarspel and Leiduin treatment plants and of samples from the distribution points Distribution 1(DN1), Distribution 2 (DN2) and Distribution 3 (DN3) for sampling date of 15th January, 2008.

Figure 3.16 shows the SEC-OCD chromatogram of finished waters of WPK and LDN treatment plants and of samples from the three distribution points, DN1, DN2 and DN3 for sampling done on the 15th January 2008. The similarity between finished water of Weesperkarspel and DN1 and between finished water of Leiduin and DN3 is apparent. The NOM composition for the sample from DN2 was intermediate between that of the two finished waters and strongly dependant on the sampling time.

Table 3.5 Mean values of DOC, SEC-OCD fractions, AOC, F-EEMs, Aeromonas spp. counts and ATP of finished waters for Weesperkarspel (WPK) and Leiduin (LDN) and of waters from the three distributions points, DN1, DN2 and DN3.

Parameter	Units	WPK	DN1	DN2	DN3	LDN
DOC	(mg C/L)	3.0	3.1	2.4	1.0	1.1
Biopolymers	(µg C/L)	25	24	21	5	5
Humics	(µg C/L)	2148	2217	1727	650	668
Building blocks	(µg C/L)	509	516	409	181	179
LMW acids	(µg C/L)	8	15	5	1.1	0.5
Neutrals	(µg C/L)	258	279	226	126	128
AOC total	(µg ac-C/L)	20.6	29.7	36.0	19.1	7.0
AOC NOX	(µg ac-C/L)	18.9	16.0	18.7	8.5	6.5
AOC P17	(µg ac-C/L)	1.7	13.8	17.3	10.5	0.5
Tyrosine-like B peak	(R.U.)	0.02	0.03	0.00	0.02	0.02
Tryptophan-like T peak	(R.U.)	0.03	0.04	0.02	0.02	0.02
Humic-like M peak	(R.U.)	0.09	0.1	0.08	0.05	0.05
Humic-like C peak	(R.U.)	0.08	0.09	0.07	0.04	0.04
Aeromonas spp.	(cfu.l^{-1})	103	288	1182	12	4
ATP	(ng.l^{-1})	2.5	2.6	2.5	1.4	1.6

Table 3.5 shows the mean value of the DOC concentration, concentration of SEC-OCD fractions, the concentration of AOC (total, NOX, and P17), peak fluorescence intensities, *Aeromonas* spp counts and ATP for finished waters of WPK and LDN water treatment plants, and of water samples from the three distribution points, DN1, DN2, and DN3. There was no significant ($p > .05$) difference in the mean concentration of DOC and SEC-OCD fractions between WPK finished water and samples from DN1 as well between LDN finished water and samples from DN3. The concentrations for DN2 samples were generally lower than for WPK but higher than for LDN. Similarly, there was no significant ($p > .05$) difference in the mean fluorescence intensity of all of the four peaks between WPK and DN1 as well as between LDN and DN3. However, whereas the mean fluorescence intensities of the protein-like peaks B and T for WPK finished water were not significantly different from those of LDN finished water, those of humic-like peaks M and C were higher for WPK finished water and DN1 than for LDN finished water and DN3. The mean fluorescence intensity of tyrosine-like peak B was not significantly ($p > .05$) between the finished waters of WPK and LDN or between the finished water and any of the distribution samples. The concentration of total AOC as well as of AOC NOX and AOC P17 in the finished waters of WPK and LDN showed a similar trend to that of DOC and SEC-OCD concentrations: the concentrations were generally about three times higher for WPK than for LDN. However, unlike DOC and

SEC-OCD concentrations, the total AOC concentrations were significantly ($p > .05$) higher in the distribution network than in the finished waters of WPK and LDN.

3.4   Conclusions

Based on the results of NOM characterization of water samples from LVN/WPK and LDN drinking water treatment trains as well as from their common water distribution network using bulk NOM measurements, F-EEMs and SEC-DOC, the following conclusions can be drawn:

- Characterization of NOM with F-EEM and SEC-DOC facilitates the identification and evaluation of the fate of NOM fractions of interest in drinking water treatment.

- F-EEM and SEC-DOC results demonstrate that both LVN/WPK and LDN treatment trains significantly reduced the aromaticity of the humic fraction in the water

- The finished water with a higher (LVN/WPK) concentration of humic substances had higher concentrations of AOC, ATP and *Aeromonas spp* than the finished water with a lower (LDN) concentration of humic substances

- In order to lower the concentration of AOC in the finished water of LVN/WPK to less than 10 µg ac-C.L^{-1}, which is the threshold for biologically stable water, further removal of humic substances prior to ozonation by an additional treatment process such as ion-exchange may be required

3.5   References

Abbt-Braun, G., Lankes, U. and Frimmel, F.H. 2004 Structural characterization of aquatic humic substances – The need for a multiple method approach. *Aquatic Sciences* 66, 151-170.

Aiken, G. and Cotsaris, E. 1995 Soil and hydrology: Their effect on NOM. *J. Am. Water Works Assoc.* 87(1), 36-45

Allpike, B.P., Heitz, A., Joll, C.A., Kagi, R.I., Abbt-Braun, G., Frimmel, F.H., Brinkmann, T., Her, N. and Amy, G. 2005 Size Exclusion Chromatography To Characterize DOC Removal in Drinking Water Treatment. *Environ. Sci. Technol.* 39(7), 2334-2342.

Allpike, B.P., Heitz, A., Joll, C.A. and Kagi, R.I. 2007 A new organic carbon detector for size exclusion chromatography. *J. Chromatogr.* A 1157 472-476.

Amy, G.L., Chadik, P.A. and Chowdhury, Z.K. 1987 Developing models for predcting THM formation potential and kinetics. *J. Am. Water Works Assoc.* 79, 89-97.

Baker, A. 2001 Fluorescence excitation-emission matrix characterization of some sewage-impacted rivers. *Environ. Sci. Technol.* 35(5), 948-953.

Bolto, B., Dixon, D., Eldridge, R. and King, S. 2002 Removal of THM precursors by coagulation or ion exchange. *Water Res.* 36(20), 5066-5073.

Chen, J., Gu, B.H., LeBoeuf, E.J., Pan, H.J. and Dai, S. 2002 Spectroscopic characterization of the structural and functional properties of natural organic matter fractions. *Chemosphere* 48(1), 59-68.

Chen, W., Westerhoff, P., Leenheer, J.A. and Booksh, K. 2003 Fluorescence Excitation-Emission Matrix Regional Integration to Quantify Spectra for Dissolved Organic Matter. *Environ. Sci. Technol.* 37, 5701-5710.

Chin, Y.-P., Aiken, G. and O'Loughlin, E. 1994 Molecular Weight, Polydispersity, and Spectroscopic Properties of Aquatic Humic Substances. *Environ. Sci. Technol.* 28, 1853-1858.

Chow, C.W.K., van Leeuwen, J.A., Drikas, M., Fabris, R., Spark, K.M. and Page, D.W. 1999 The impact of the character of natural organic matter in conventional treatment with alum. *Water Sci. Technol.* 40(9), 97-104.

Coble, P.G., Green, S.A., Blough, N.V. and Gagosian, R.B. 1990 Characterization of dissolved organic matter in the Black Sea by fluorescence spectroscopy. *Nature* 348, 432-435.

Coble, P.G., Schultz, C.A. and Mopper, K. 1993 Fluorescence contouring analysis of DOC Intercalibration Experiment samples: a comparison of techniques. *Marine Chemistry* 41, 173-178.

Coble, P.G. 1996 Characterization of marine and terrestrial DOM in seawater using excitation-emission matrix spectroscopy. *Marine Chemistry* 51, 325-346.

Croué, J.-P., G.V.Korshin and M.M.Benjamin (eds) (2000) Characterization of Natural Organic Matter in Drinking Water, AwwRF, Denver, CO.

Delahaye, E., Welte, B., Levi, Y., Leblon, G. and Montiel, A. 2003 An ATP-based method for monitoring the microbiological drinking water quality in a distribution network. *Water Res.* 37(15), 3689-3696.

Edzwald, J.K., Becker, W.C. and Wattier, K.L. 1985 Surrogate parameter for monitoring organic matter and THM precursors. *J. Am. Water Works Assoc.* 77, 122-132.

Frimmel, F.H., Gremm, T. and Huber, S. 1992 Liquid-Chromatographic Characterization of Refractory Organic-Acids. *Sci. Total Environ.* 118, 197-206.

Frimmel, F.H. 1998 Characterization of natural organic matter as major constituents in aquatic systems. *Journal of Contaminant Hydrology* 35, 201–216.

Frimmel, F.H. and Abbt-Braun, G. 1999 Basic Characterization of Reference NOM from Central Europe - Similarities and Differences. *Environment International* 25(2/3), 191-207.

Havelaar, A.H., During, M. and Versteegh, J.F.M. 1987 Ampicillin-dextrin agar medium for the enumeration of Aeromonas species in water by membrane filtration. *J. Appl. Bacteriol.* 62, 279-287.

Hem, L.J. and Efraimsen, H. 2001 Assimilable organic carbon in molecular weight fractions of natural organic matter. *Water Res.* 35(4), 1106-1110.

Her, N., Amy, G., McKnight, D., Sohna, J. and Yoon, Y. 2003 Characterization of DOM as a function of MW by fluorescence EEM and HPLC-SEC using UVA, DOC, and fluorescence detection. *Water Res.* 37, 4295–4303.

Huber, S.A., Balz, A., Abert, M. and Pronk, W. 2011 Characterisation of aquatic humic and non-humic matter with size-exclusion chromatography – organic carbon detection – organic nitrogen detection (LC-OCD-OND). *Water Res.* 45, 879-885.

Hwang, C., Krasner, S., Sclimenti, M., Amy, G. and Dickenson, E. (eds) (2002) Polar NOM: characterization, DBPs, treatment American Water Works Association Research Foundation, Denver, CO.

Jaffe', R., McKnight, D., Maie, N., Cory, R., McDowell, W.H. and Campbell, J.L. 2008 Spatial and temporal variations in DOM composition in ecosystems: The importance of long-term monitoring of optical properties. *Journal of Geophysical Research* 113(G04032, doi:10.1029/2008JG000683).

Korshin, G., Chow, C.W.K., Fabris, R. and Drikas, M. 2009 Absorbance spectroscopy-based examination of effects of coagulation on the reactivity of fractions of natural organic matter with varying apparent molecular weights. *Water Res.* 43, 1541-1548.

Lee, N., Amy, G. and Croue, J.-P. 2006 Low-pressure membrane (MF/UF) fouling associated with allochthonous versus autochthonous natural organic matter. *Water Res.* 40, 2357 – 2368.

Leenheer, J.A., Croué, J.-P., Benjamin, M., Korshin, G.V., Hwang, C.J., Bruchet, A. and Aiken, G.R. (2000) *Comprehensive Isolation of Natural Organic Matter from Water for Spectral Characterizations and Reactivity Testing*. In: Natural Organic Matter and Disinfection By-Products. American Chemical Society, pp. 68-83.

Leenheer, J.A. and Croue, J.-P. 2003 Characterizing Dissolved Aquatic Organic matter: Understanding the unknown structures is key to better treatment of drinking water. *Environ. Sci. Technol.* 37(1), 19A-26A.

Leenheer, J.A. 2004 Comprehensive assessment of precursors, diagenesis, and reactivity to water treatment of dissolved and colloidal organic matter. *Water Sci. Technol. Water Supply* 4(4), 1-9.

Magic-Knezev, A. and van der Kooij, D. 2004 Optimisation and significance of ATP analysis for measuring active biomass in granular activated carbon filters used in water treatment. *Water Research* 38(18), 3971-3979.

Mopper, K. and Schultz, C.A. 1993 Fluorescence as a possible tool for studying the nature and water column distribution of DOC components. *Marine Chemistry* 41, 229-238.

Nam, S.N. and Amy, G. 2008 Differentiation of wastewater effluent organic matter (EfOM) from natural organic matter (NOM) using multiple analytical techniques. *Water Sci. Technol.* 57(7), 1009-1015.

Owen, D.M., Amy, G.L., Chowdhury, Z.K., Paode, R., McCoy, G. and Viscosil, K. 1998 NOM characterization and treatability *J. Am. Water Works Assoc.* 87(1), 46-63.

Peuravuori, J. and Pihlaja, K. 1997 Molecular size distribution and spectroscopic properties of aquatic humic substances. *Analytica Chimica Acta* 337, 133-149.

Peuravuori, J., Koivikko, R. and Pihlaja, K. 2002 Characterization, differentiation and classification of aquatic humic matter separated with different sorbents: synchronous scanning fluorescence spectroscopy. *Water Res.* 36, 4552–4562.

Reemtsma, T., These, A., Springer, A. and Linscheid, M. 2008 Differences in the molecular composition of fulvic acid size fractions detected by size-exclusion chromatography–on line Fourier transform ion cyclotron resonance (FTICR–) mass spectrometry. *Water Res.* 42, 63-72.

Senesi, N., Miano, T.M., Provenzano, M.C. and Brunetti, G. 1989 Spectroscopic and compositional characterization of I.H.S.S. reference and standard fulvic and humic acids of various origin. *Sci. Total Environ.* 81(2), 143-156.

Stedmon, C.A. and Markager, S. 2005 Resolving the variability in dissolved organic matter fluorescence in a temperate estuary and its catchment using PARAFAC analysis. *Limnol. Oceanogr.* 50(2), 686-697.

Stedmon, C.A., Seredynska-Sobecka, B., Boe-Hansen, R., Tallec, N.L., Waul, C.K. and Arvin, E. 2011 A potential approach for monitoring drinking water quality from groundwater systems using organic matter fluorescence as an early warning for contamination events. *Water Res.* 45(18), 6030-6038.

Swietlik, J., Dabrowska, A., Raczyk-Stanislawiak, U. and Nawrocki, J. 2004 Reactivity of natural organic matter fractions with chlorine dioxide and ozone. *Water Res.* 38(3), 547-558.

Traina, S.J., Novak, J. and Smeck, N.E. 1990 An Ultraviolet Absorbance Method of Estimating the Percent Aromatic Carbon Content of Humic Acids. *J. Environ. Qual.* 19(1), 151-153.

van der Kooij, D. 1992 Assimilable organic carbon as an indicator of bacterial regrowth. *J. Am. Water Works Assoc.* 84(2), 57-65.

World Health Organisation (WHO) (2003) Managing regrowth in drinking water distribution systems. Bartram, J., Cotruvo, J., Exner, M., Fricker, C. and Glasmacher, A. (eds).

Volk, C., Bell, K., Ibrahim, E., Verges, D., Amy, G. and Lechevallier, M. 2000 Impact of Enhanced and Optimized Coagulation on Removal of Organic Matter and its Biodegradable Fraction in Drinking Water. *Water Res.* 34(12), 3247-3257.

Vuorio, E., Vahala, R., Rintala, J. and Laukkanen, R. 1998 The evaluation of drinking water treatment performed with HPSEC. *Environment International* 24(5/6), 617-623.

Weishaar, J.L. 2003 Evaluation of Specific Ultraviolet Absorbance as an Indicator of the Chemical Composition and Reactivity of Dissolved Organic Carbon. *Environ. Sci. Technol.* 37, 4702-4708.

Wong, S., Hanna, J.V., King, S., Carroll, T.J., Eldridge, R.J., Dixon, D.R., Bolto, B.A., Hesse, S., Abbt-Braun, G. and Frimmel, F.H. 2002 Fractionation of natural organic matter in drinking water and characterization by C-13 cross-polarization magic-angle spinning NMR spectroscopy and size exclusion chromatography. *Environ. Sci. Technol.* 36(16), 3497-3503.

Wu, F.C., Evans, R.D. and Dillon, P.J. 2003 Separation and Characterization of NOM by High-Performance Liquid Chromatography and On-Line Three-Dimensional Excitation Emission Matrix Fluorescence Detection. *Environ. Sci. Technol.* 37, 3687-3693.

Yamashita, Y. and Tanoue, E. 2003 Chemical characterization of protein-like fluorophores in DOM in relation to aromatic amino acids. *Marine Chemistry* 82, 255-271.

Chapter 4

TRACKING NATURAL ORGANIC MATTER (NOM) IN A DRINKING WATER TREATMENT PLANT USING FLUORESCENCE EXCITATION–EMISSION MATRICES AND PARAFAC

A part of this chapter has been published as:

Baghoth, S.A., Sharma, S.K. and Amy, G.L. 2011 Tracking natural organic matter (NOM) in a drinking water treatment plant using fluorescence excitation-emission matrices and PARAFAC. *Water Res.* 45(2), 797-809.

Summary

Natural organic matter (NOM) in water samples from two drinking water treatment trains was characterized using fluorescence excitation emission matrices (F-EEMs) and parallel factor analysis (PARAFAC). A seven component PARAFAC model was developed and validated using 147 F-EEMs of water samples from two full-scale water treatment plants. The fluorescent components have spectral features similar to those previously extracted from F-EEMs of dissolved organic matter (DOM) from diverse aquatic environments. Five of these components are humic-like with a terrestrial, anthropogenic or marine origin, while two are protein-like with fluorescence spectra similar to those of tryptophan-like and tyrosine-like fluorophores. A correlation analysis was carried out for samples of one treatment plant between the maximum fluorescence intensities (F_{max}) of the seven PARAFAC components and NOM fractions (humics, building blocks, neutrals, biopolymers and low molecular weight acids) of the same sample obtained using size exclusion chromatography with organic carbon detection (SEC-OCD). There were significant correlations ($p < 0.01$) between sample DOC concentration, UVA_{254}, and F_{max} for the seven PARAFAC components and DOC concentrations of the SEC-OCD fractions. Three of the humic-like components showed slightly better predictions of DOC and humic fraction concentrations than did UVA_{254}. Tryptophan-like and tyrosine-like components correlated positively with the biopolymer fraction. These results demonstrate that fluorescent components extracted from F-EEMs using PARAFAC could be related to previously defined NOM fractions and that they could provide an alternative tool for evaluating the removal of NOM fractions of interest during water treatment.

4.1   Introduction

Natural organic matter (NOM) is a major concern in drinking water treatment since it causes adverse aesthetic qualities such as colour, taste and odour. It negatively affects the performance of water treatment processes such as granular activated carbon filtration and membrane filtration and it may promote biogrowth in water distribution networks. Furthermore, it can decrease the effectiveness of oxidants and disinfectants and produce undesirable disinfection by-products during oxidation processes (Owen et al., 1993). Thus, in order to minimise these undesirable effects, it is essential to limit the concentration of NOM in treated water. However, both the amount and composition of NOM affect the efficiency of its removal during water treatment. Therefore, in order to design and operate drinking water treatment processes for optimal NOM removal, a better understanding of its character is necessary.

Many studies and reviews have been undertaken on the structural characterization of aquatic NOM (Frimmel, 1998; Abbt-Braun et al., 2004; Leenheer, 2004) but its structure and fate in drinking water treatment (individual processes and process trains) are still not fully understood. Because NOM may contain literally thousands of different chemical constituents, it is not realistic to characterize it on the basis of a thorough compilation of the individual compounds (Croué et al., 2000). Therefore, researchers have found it more practical to characterize NOM according to operationally defined chemical groups having similar properties. These groups are commonly isolated by methods which involve concentration and fractionation of bulk NOM (Frimmel and Abbt-Braun, 1999; Peuravuori et al., 2002). Whereas these methods provide valuable insight into the nature of NOM from diverse aquatic environments, they are often laborious, time consuming and may involve extensive pre-treatment of samples which could modify the NOM character. Thus, they are not commonly used for monitoring of NOM in drinking water treatment plants.

A technique that has recently gained popularity for NOM characterization is fluorescence spectroscopy. It is a simple, relatively inexpensive and very sensitive tool that requires little or no sample pre-treatment. Since fluorescence is a function of structure and functional groups in molecules, it can be used to extract a lot of information about the chemical characteristics of NOM. Hudson et al. (2007) carried out an extensive review of the use of fluorescence spectroscopy to measure organic matter fluorescence and the application of dissolved organic matter (DOM) fluorescence in marine waters, freshwaters and wastewaters. They concluded that whereas the investigation of the source, character and reactions of marine organic matter is common, the investigation of the behaviour of organic matter in freshwaters still lags behind marine waters.

Three-dimensional fluorescence excitation emission matrix (F-EEM) spectroscopy has been used to distinguish different types and sources of dissolved organic carbon (DOC) in natural waters (Coble et al., 1990). It has been used to characterize DOC and to identify humic-like and protein-like fluorescent signals in water samples from different aquatic environments (Coble, 1996). In a study of sewage impacted rivers using F-EEM spectroscopy, protein fluorescence was found to be a better indicator of sewage pollution than ultraviolet (UV) absorbance at 254 nm (UVA$_{254}$) (Baker, 2001).

Various methods have been used to analyze F-EEMs. The traditional peak picking method involves the use of excitation-emission wavelength pairs to identify fluorophores based on the location of the maximum fluorescence intensity (Coble, 1996). The fluorescence intensity peaks are picked from a contour plot of F-EEMs and the excitation and emission wavelength pairs at which they occur are used to characterize the NOM fluorescence. A review of recent literature demonstrated the potential of F-EEMs as a successful monitoring tool for recycled water systems (Henderson et al., 2009 ; Bieroza et al., 2009a) used F-EEMs for the assessment of TOC removal and organic matter characterization of surface waters and they found that F-EEMs could be used to predict TOC removal during surface water treatment by clarification. F-EEMs have also been used to distinguish between allochthonous and autochthonous DOC on the basis of a fluorescence index (FI), which is calculated as a ratio of fluorescence intensity at emission wavelength of 450 nm to that at 500 nm, obtained with an excitation wavelength of 370 nm (McKnight et al., 2001).

More recent methods for the analysis of DOM EEMs include fluorescence regional integration (FRI) (Chen et al., 2003), multivariate data analysis (e.g. Principal Component Analysis, PCA, and Partial Least Squares regression, PLS) (Persson and Wedborg, 2001), and multi-way data analysis using parallel factor analysis (PARAFAC) (Stedmon et al., 2003). PCA of F-EEMs has been used to identify major foulants for ultrafiltration (UF) and nanofiltration (NF) membranes and to assess the performance of feed water pre-treatment by roughing filters, biofilters and the subsequent UF/NF membrane filters (Peiris et al., 2010). Recently, F-EEMs have been used with self-organising maps for determination of NOM removal efficiency in water treatment works (Bieroza et al., 2009b).

More detailed information about NOM character of water samples can be obtained by using F-EEMs and PARAFAC, a statistical method used to decompose multi-dimensional data. F-

EEMs may be arranged in three dimensions comprising fluorescence measurements at several excitation and emission wavelengths for several samples and the resulting three-way data modelled with PARAFAC. In this way, individual components have been extracted some of which have been attributed to protein-like, fulvic-like or humic-like fractions of NOM. Although the method was first used for NOM characterization only recently (Stedmon et al., 2003), it has been used in several studies of DOM (Stedmon and Markager, 2005a; Hunt and Ohno, 2007; Yamashita and Jaffe, 2008).

In a study of DOM from a wide variety of aquatic environments, F-EEMs and PARAFAC were used to identify thirteen components, seven of which were attributed to quinone-like fluorophores (Cory and McKnight, 2005). However, unlike all the other studies involving PARAFAC analysis of DOM fluorescence, which used oxidized materials only, Cory and McKnight, 2005) used reduced and oxidized samples, thus resulting in the extraction of more components in their study.

Another property which is important for understanding the physical and chemical characteristics of NOM is molecular size (MS) or molecular weight (MW). It influences the adsorption, bioavailability as well as other water treatment processes for the removal of NOM. Lower MW NOM molecules tend to be more hydrophilic and thus more biolabile, while higher MW NOM molecules tend to be more aromatic and more hydrophobic, and have higher affinity for adsorption. High performance size exclusion chromatography (HPSEC), which separates molecules according to molecular size or molecular weight, has been widely applied in characterization of NOM in aquatic environments (Chin et al., 1994; Her et al., 2003; Croué, 2004). It has been shown to be very effective in following changes in the NOM distribution along drinking water treatment trains (Vuorio et al., 1998; Matilainen et al., 2002).

HPSEC may be coupled with detectors such as UV, fluorescence or DOC detectors. An HPSEC system coupled with an organic carbon detector (SEC-OCD) (Huber and Frimmel, 1994) has been used to fractionate NOM into five fractions: biopolymers (such as polysaccharides, polypeptides, proteins and amino sugars); humic substances (fulvic and humic acids); building blocks (hydrolysates of humic substances); low molecular weight (LMW) humic substances and acids; and low molecular weight neutrals (such as alcohols, aldehydes, ketones and amino acids).

HPSEC has been used with online F-EEM to study fluorescence properties of NOM as a function of MS and polarity (Wu et al., 2003; Her et al., 2003). Allpike et al., 2005 used HPSEC with online DOC, UV and fluorescence detectors to compare the removal of different molecular weights of DOC in two water treatment processes. Whereas Allpike et al., 2005) and Her et al., 2003) used a single pair of ex/em wavelengths for the fluorescence measurements, Wu et al., 2003) used three-dimensional F-EEM. In this study, the SEC-OCD system used was coupled with online UV and DOC detectors and the F-EEMs were measured separately.

As well as contributing to a better understanding of NOM, identification of fluorescent components using PARAFAC could be used to track the fate of problematic NOM fractions and to optimise the design and operation of drinking water treatment processes for their removal. Recently, fluorescence spectroscopy has been used for organic matter characterization and assessment of TOC removal in drinking water treatment (Bieroza et al., 2009a; Bieroza et al., 2010). However, they used F-EEMs alone, which were collected from fewer treatment processes, while this study uses F-EEMs and SEC-OCD. Furthermore, they used fluorescence intensity peaks identified from the composite EEMs, while this study uses PARAFAC to identify fluorescence intensity peaks for individual fluorescent components. The main objective of this study was to characterize NOM in samples from a drinking water treatment train using F-EEMs and PARAFAC. A further objective was to examine the relationship between the extracted PARAFAC components and the NOM fractions of the same samples obtained using SEC-OCD.

4.2   Methods

4.2.1 Sampling

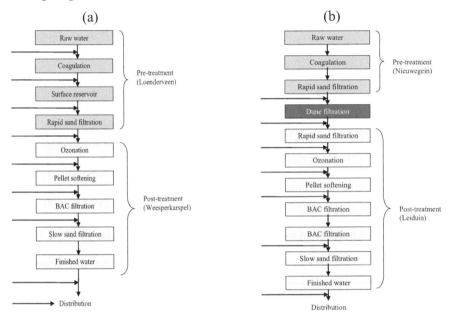

Figure 4.1 Water treatment scheme and sampling points (arrows) for the treatment process trains of (a) Loenderveen/Weesperkarspel, and (b) Leiduin.

Water samples were collected monthly from different points along the process trains (Figure 4.1) of two drinking water treatment plants, operated by Waternet, which supply water to Amsterdam city and its environs. Samples were collected between January and December 2007 from Loenderveen/ Weesperkarspel treatment train consisting of two stages: a pre-treatment plant at Loenderveen, which treats surface water by coagulation and flocculation, retention in surface water reservoir for about 100 days and rapid sand (RS) filtration; and a post treatment plant at Weesperkarspel, about 10 km away, which comprises ozonation, pellet softening, biological activated carbon (BAC) filtration and slow sand (SS) filtration. On average, twelve samples were collected from Loenderveen/ Weesperkarspel process train and from the distribution network every month. Samples were also collected from June to September in 2008 and in August and September of 2009 from the other treatment train, at Leiduin, which treats surface water which has been pre-treated by coagulation, rapid sand filtration at Nieuwegein, and then by infiltration in sand dunes. The Leiduin process train comprises aeration, RS filtration, ozonation, softening, BAC filtration and SS filtration.

Seven samples were collected each time. The samples were collected in clean glass bottles and then filtered through pre-washed 0.45 μm regenerated cellulose membrane filters within 24 hours of arrival in the laboratory. The filtered samples were then stored at 5°C until analysis, which was done within one week. The following measurements were performed for the samples: DOC, UV absorbance at 254 nm (UVA$_{254}$), F-EEM and SEC-OCD.

4.2.2   DOC and UV$_{254}$ measurements

DOC concentrations of all pre-filtered samples were determined by the combustion method using a Shimadzu TOC-V$_{CPN}$ organic carbon analyzer. UVA$_{254}$ absorbance of each sample was measured in a 1 cm quartz cell using a Shimadzu UV-2501PC UV-VIS spectrophotometer. For each sample, SUVA was determined by dividing the absorbance UVA$_{254}$ by the corresponding DOC concentration.

4.2.3   Fluorescence EEM measurements

To minimise fluorescence quenching resulting from the relatively high concentrations of DOC (inner filter effects), the pre-filtered samples were diluted to a DOC concentration of 1 mg C/L using 0.01 M KCl solution prior to fluorescence measurements. To minimise possible metal-NOM complexation, the pH of diluted samples was adjusted to 2.8 ± 0.1 using 0.1 M HCl and the fluorescence intensities measured in a 1.0 cm quartz cell using a FluoroMax-3 spectrofluorometer (Horiba Jobin Yvon) at room temperature (20±1°C). EEMs were generated for each sample by scanning over excitation wavelengths between 240 and 450 nm at intervals of 10 nm and emission wavelengths between 290 and 500 nm at intervals of 2 nm. The bandwidths on excitation and emission modes were both set at 1 nm. An EEM of the 0.01M KCl solution was obtained and subtracted from the EEM of each sample in order to remove most of the Raman scatter peaks. Since samples were previously diluted to a DOC concentration of 1 mg C/L, each blank subtracted EEM was multiplied by the respective dilution factor and Raman-normalized by dividing by the integrated area under the Raman scatter peak (excitation wavelength of 350 nm) of the corresponding Milli-Q water and the fluorescence intensities reported in Raman units (RU).

4.2.4   PARAFAC modelling

PARAFAC was used to model the dataset of F-EEMs. It uses an alternating least squares algorithm to minimize the sum of squared residuals in a trilinear model, thus allowing the estimation of the true underlying EEM spectra (Bro, 1997). It reduces a dataset of EEMs into a set of trilinear terms and a residual array (Andersen and Bro, 2003):

$$x_{ijk} = \sum_{f=1}^{F} a_{if}b_{jf}c_{kf} + \varepsilon_{ijk} \qquad i = 1,...,I; \; j = 1,...,J; \; k = 1,...,K$$

where x_{ijk} is the fluorescence intensity of the ith sample at the kth excitation and jth emission wavelength; a_{if} is directly proportional to the concentration of the fth fluorophore in the ith sample (defined as scores), b_{jf} and c_{kf} are estimates of the emission and excitation spectra respectively for the fth fluorophore (defined as loadings), F is the number of fluorophores (components) and ε_{ijk} is the residual element, representing the unexplained variation in the model (Stedmon et al. 2003).

Some components extracted by PARAFAC can be ascribed to specific components of organic matter present in water samples, but they more likely represent groups of organic compounds having similar fluorescence properties. While component scores indicate the relative concentrations of groups of organic fractions represented by the components, excitation and emission loadings indicate their characteristic excitation and emission spectra. However, since most of the components that have been extracted from aquatic samples thus far cannot be ascribed to specific organic compounds, the scores cannot be converted to concentrations. Nevertheless, differences in component scores can be used to illustrate variations in the organic matter composition of water samples within a given dataset. But it should be noted that these differences may also be due to changes in the local environment of the analyte, such as polarity and temperature. In this study, differences in scores due to solution environment were minimised by performing fluorescence measurements at the same pH (2.8 ±0.1) and temperature ($20\pm1^{\circ}C$). The maximum fluorescence intensity for each component was obtained and used to illustrate the quantitative and qualitative differences between samples.

Several diagnostic tools can be used to determine the appropriate number of PARAFAC components. In this study, however, only two methods were mainly employed: split-half analysis (Harshman, 1984) and examination of residual error plots (Stedmon and Bro, 2008). For split-half analysis, the data were split in the first mode comprising of water samples. The samples were divided into two halves and a PARAFAC model obtained for each half. The excitation and emission spectral loadings of the two halves were then compared to ascertain whether they were similar.

A series of PARAFAC models consisting of between three and seven components were generated using the DOMfluor toolbox (Stedmon and Bro, 2008), which was specifically developed to perform PARAFAC analysis of DOM fluorescence, and contains all of the tools used to identify outlier samples as well as to perform split-half and residual errors diagnostics. A dataset of F-EEMs for 137 samples collected from Loenderveen /Weesperkarspel water treatment train over 12 months in 2007, and for 46 samples collected from Leiduin water treatment train during two campaigns in 2008 and 2009 was used to develop the PARAFAC model.

4.2.5 Size exclusion chromatography with organic carbon detection (SEC-OCD)

NOM separation by size exclusion was performed with an SEC-OCD system (DOC-LABOR, Germany) at Het Waterlaboratorium, Haarlem, The Netherlands. In the system, a column TSK HW-50S is connected to a Graentzel thin-film reactor (Huber and Frimmel, 1994) in which NOM is oxidized to CO_2 by UV before it is measured by infrared detection. The column separates NOM, according to molecular size/weight, up to five fractions: (i) biopolymers (BP), comprising polysaccharides, proteins and colloids, (ii) humic substances (HS), (iii) building blocks (hydrolysates of humics) (BB), (iv) low molecular weight humics and acids (LMW), and (v) low molecular weight neutrals (such as alcohols, aldehydes, ketones and amino acids). Besides the organic carbon detector, the system also incorporates a UV detector, which may be used to assess the aromaticity of the sample as well as of the humic fraction by computing the respective SUVA values, and a dissolved organic nitrogen (DON) detector. Water samples were analyzed without any pre-treatment other than filtration through 0.45 mm filters prior to injection in the chromatographic column. The classification of SEC-OCD fractions is based on empirical as well as systematic studies. For identification of HS fraction, up to five criteria may be used: (i) retention time, (ii) peak width, (iii) peak symmetry, (iv) the ratio of the peak area for the UV signal to that of the peak area for the DOC signal, and (v) DON. Definition of the fraction boundaries and quantification of the fractions by area integration of chromatograms was done with FIFFIKUS software (DOC-LABOR), which uses data for calibration standards as some of the inputs. The SEC-OCD results are presented and discussed in chapter 3.

4.2.6 Correlation analysis

Spearman rank-order correlation coefficients were used to investigate the relationships between sample DOC, UVA_{254}, maximum fluorescence intensity of the PARAFAC components (F_{max}), and the five SEC-OCD fractions (biopolymers, humics, building blocks, low molecular acids, and low molecular weight neutrals). The analysis was performed with SPSS statistical software.

4.3   Results and discussion

4.3.1 DOC, UVA_{254} and SUVA

Table 4.1 shows a summary of the means and standard deviations of the pH, DOC and SUVA of water samples collected from the Loenderveen/Weesperkarspel and Leiduin drinking water treatment trains. There were some slight seasonal variations in the DOC of the raw water at the pre-treatment plant at Loenderveen, with a minimum of 7.6 mg C/L in autumn, a maximum of 9.8 mg C/L in winter and a monthly average of 9.0 mg C/L. The retention in a surface reservoir for a period of about 100 days dampens the seasonal variation and after RS filtration, the monthly average DOC is 6.0 mg C/L. At the final treatment plant of Weesperkarspel, the DOC is further reduced by 55% to 2.7 mg C/L. The DOC levels in Leiduin were generally lower, with a mean of 2.5 mg C/L for influent pre-treated water, and a mean of 1.0 mg C/L for treated water, representing a total reduction of 60% across the process train.

Table 4.1 Variation of DOC concentrations and SUVA for samples from Loenderveen/ Weesperkarspel and Leiduin drinking water treatment trains.

Loenderveen/Weesperkarspel process train				Leiduin process train			
Sample	pH	DOC[a] (mg C/L)	SUVA[a] (L/mg/m)	Sample	pH	DOC[a] (mg C/L)	SUVA[a] (L/mg/m)
Raw water	7.9 ± 0.3	9.0 ± 0.8	3.5 ± 0.3	Pretreated water	8.0 ± 0.1	2.5 ± 0.4	2.7 ± 0.4
Coagulation effluent	7.7 ± 0.3	7.1 ± 0.6	3.0 ± 0.1	RS filtration effluent	7.9 ± 0.2	2.1 ± 0.3	2.6 ± 0.2
Surface reservoir effluent	7.8 ± 0.3	6.5 ± 0.2	3.0 ± 0.1	Ozonation effluent	8.0 ± 0.1	2.0 ± 0.2	1.8 ± 0.2
RS filtration effluent	7.8 ± 0.3	6.0 ± 0.3	2.8 ± 0.1	Pellet softening effluent	7.9 ± 0.1	1.9 ± 0.3	1.7 ± 0.2
Ozonation effluent	7.8 ± 0.3	5.7 ± 0.3	1.8 ± 0.1	BAC filtration effluent	8.1 ± 0.3	1.2 ± 0.2	1.2 ± 0.1
Pellet softening effluent	7.9 ± 0.3	5.4 ± 0.3	1.7 ± 0.1	Finished water	8.2 ± 0.2	1.0 ± 0.1	1.2 ± 0.2
BAC filtration effluent	8.1 ± 0.2	3.0 ± 0.5	1.5 ± 0.1				
Finished water	8.1 ± 0.2	2.7 ± 0.3	1.5 ± 0.1				

[a] Mean value ± standard deviation, for n = 13 and 7 for Loenderveen/Weesperkarspel and Leiduin, respectively.

The average SUVA for raw water of Loenderveen/ Weesperkarspel treatment train was 3.5 L/mg/m, an indication of NOM of moderate aromaticity, while that of finished water was 1.5 L/mg/m, which is typical of NOM with low aromaticity (SUVA<2 L/mg/m). For Leiduin treatment train, the SUVA for the influent water, previously pre-treated by coagulation and filtration followed by infiltration in sand dunes, was 2.7 L/mg/m, while that of finished water was 1.2 L/mg/m. Thus, both Loenderveen/Weesperkarspel and Leiduin treatment trains significantly reduce the aromatic character of the NOM in the treated water.

4.3.2 PARAFAC components

A total of 183 F-EEMs of water samples from Loenderveen/Weesperkarspel and Leiduin water treatment trains were used for PARAFAC analysis. An initial exploratory analysis was performed in which outliers were identified and removed from the dataset. A sample was considered an outlier if it contained some instrument error or artefact or if it was properly measured but was very different from the others (determined by calculating its leverage using DOMfluor). The latter was removed in order to facilitate the modelling process as well as the model validation using the split-half method; otherwise, the dataset would need to contain a sufficient number of the unique samples, which are evenly divided between the split halves.

Table 4.2 Comparison of the spectral characteristics of the seven components identified in this study with those of similar components from previous studies. Values in brackets represent secondary peaks or shoulders.

Component of this study	Excitation/Emission wavelength	Description and source assignment (References)
C1	260(360)/480	Terrestrial humic substances Peak P3: <260(380)/498, (Ref.3) Component 3: 270(360)/478, (Ref.4)
C2	250(320)/410	Terrestrial/anthropogenic humic substances Component 6: <250(320)/400, (Ref.5) Component C2: 315/418, (Ref.2)
C3	<250(330)/420	Marine and terrestrial humic substances Peak M, Coble (Ref.1) Component P1: (<260)310/414, (Ref.3)
C4	<250(290)/360	Amino acids, free or protein bound Component C7: 240(300)/338, (Ref.3) Component 4: <260(305)/378, (Ref.7)
C5	250(340)/440	Terrestrial humic substances Component P8: <260(355)/434, (Ref.3) Component 4: 250(360)/440, (Ref.5)
C6	(<250)300/406	Marine and terrestrial humic substances Component 1: (<260)305/428, (Ref.7) Component 3:295/398, (Ref.6) Peak C or M: (Ref.1)
C7	270/306	Amino acids, free or protein bound Component 4: 275/306, (Ref.6) Component 8: 275/304, (Ref.5) Peak B: 275/310, (Ref.1)

1. Coble, 1996, 2. Murphy et al., 2006, 3. Murphy et al., 2008, 4. Stedmon et al. 2003, 5. Stedmon and Markager, 2005a, 6. Stedmon and Markager, 2005b, 7. Yamashita and Jaffe, 2008.

PARAFAC analysis with 3 to 7 components was performed on the remaining 147 samples. However, only the models containing three, four and seven components could be split-half validated. These were split-half validated in the sense that the corresponding components in the split halves had equal excitation and emission loadings as verified by the corresponding

Tucker's congruence coefficients being greater than 0.95 (Lorenzo-Seva and Ten Berge, 2006). For a complete dataset model to be validated, the Tucker's congruence coefficients between the split halves, as well as between the complete dataset and a split half should be greater than 0.95 and only the seven component model could be validated in this manner.

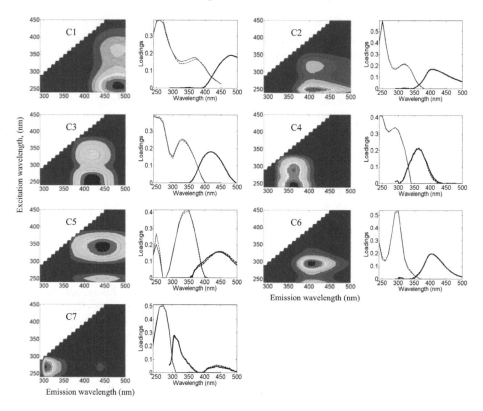

Figure 4.24 Contour plots of the seven components identified from the complete F-EEMs dataset. The line plots on the right show split-half validations of excitation (thin) and emission (thick) loadings between the complete dataset (solid) and one of the independent halves (dotted).

Whereas the PARAFAC model in this study uses F-EEMs of samples collected at a low pH (~ 2.9), which would inevitably result in the non-uniform quenching of fluorescent peaks of the different components, the seven components extracted have spectral features similar to those previously extracted from F-EEMs of DOM (Stedmon et al., 2007; Murphy et al., 2008; Borisover et al., 2009), all of which were collected at ambient pH (normally above 7.0). Table 4.2 shows excitation and emission wavelength pairs of the main peaks of the seven components as well as descriptions of similar components that were identified in previous studies. Comparison of previously identified components with the spectral contours shown in Figure 4.2 indicates that the samples in this study contain humic-like as well as protein-like fluorophores. Two of the components (C4 and C7) have previously been ascribed to protein-like fluorophores (Cory and McKnight, 2005): component C4 to tryptophan-like fluorophore, and component C7 to tyrosine-like fluorophore (Yamashita and Tanoue, 2003). Components C1, C2, C3, C5 and C6 are humic-like fluorophores which may have a terrestrial or anthropogenic origin. Component C3 was initially thought to originate only from marine environments but has now been found to be present in freshwaters influenced by agricultural inputs (Stedmon and Markager, 2005).

4.3.3 PARAFAC component scores across treatment

After validation of the seven component model, the fate of the components across the Loenderveen/Weesperkarspel treatment train was tracked using their maximum fluorescence intensities (F_{max}). F_{max} gives estimates of the relative concentrations of each component; however, direct comparison of relative concentrations between different components depends on the magnitude of their quantum efficiencies as well as on their individual responses to quenching effects. Figure 4.3 shows the mean F_{max} of each component across the treatment train. For all the water samples analyzed, F_{max} was higher for terrestrial humic-like components C1 and C2 than for humic-like components C3, C5 and C6, and for protein-like components C4 and C7. For raw water samples, the mean F_{max} was: 1.63 and 1.64 RU. for C1 and C2, respectively; 0.50, 0.42 and 0.39 for humic components C3, C5 and C6, respectively; and 0.57 and 0.25 for protein-like components C4 and C7, respectively. F_{max} of the tyrosine-like component C7 was almost always lower than that of any of the other components, while that of the tryptophan-like component C4 was comparable to those of humic-like components C3, C5 and C6. Whereas these results appear to indicate that the samples were dominated by humic-like fluorescent compounds, they are not sufficient to permit conclusions to be drawn about the relative concentrations of all the seven components without prior knowledge of their respective quantum yields. Since fluorescence intensity is proportional to concentration as well as quantum yield of the fluorophores, differences in the relative fluorescent intensities of the components may be a reflection of differences in concentrations and/ or quantum efficiencies of the components. However, results of NOM characterization of the same set of samples using SEC-OCD showed quantitatively that, on average, humic substances comprised about 70% of all samples analyzed (see chapter 3).

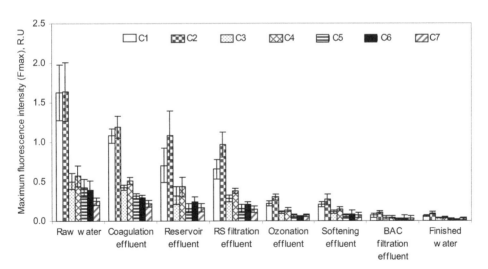

Figure 4.25 Maximum fluorescence intensities (F_{max}) of PARAFAC components across Loenderveen/ Weesperkarspel drinking water treatment train.

In order to evaluate the effect of water treatment on fluorescence characteristics of NOM, the mean percentage reduction of F_{max} across each treatment process (Figure 4.4) was computed. F_{max} may be reduced in two ways: (i) intact removal of fluorescent compounds by, for example, coagulation and BAC filtration or (ii) transformation of fluorescent compounds by, for example, ozonation. Ozonation and BAC filtration reduced F_{max} by 50% or more for all components, while coagulation and storage in surface reservoir reduced it by between 5 and 50%, respectively, depending on the component. For all components, softening and slow sand filtration did not reduce F_{max} while rapid sand filtration reduced it by less than 10%. Whereas the mean percentage reduction of fluorescence is comparable to that of DOC in the case of BAC filtration (~50-70% for fluorescence and ~ 40% for DOC), it is disproportionately higher than that of DOC in the case of ozonation (~ 50-70% for fluorescence and 5% for DOC). This is explained by the fact that ozonation transforms large molecular weight NOM into smaller and less aromatic organic compounds (Swietlik et al., 2004) which have lower UV absorptivities and fluorescence. Coagulation significantly reduced F_{max} of all humic-like components as well as of the tyrosine-like component C7 but not of the tryptophan-like component C4. The reduction in humic-like NOM is consistent with a previous study (Allpike et al., 2005) showing effective removal of larger molecular weight, hydrophobic humic-like NOM by coagulation.

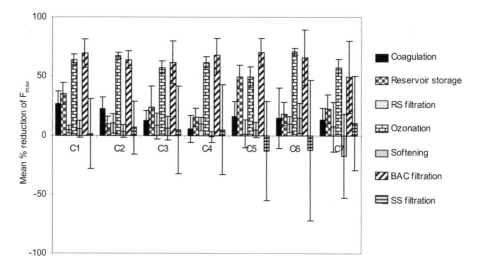

Figure 4.26 Mean percentage reduction of maximum fluorescence intensities (F_{max}) of PARAFAC components across Loenderveen/ Weesperkarspel water treatment train.

The ratios of F_{max} of components, particularly those of humic-like to protein-like, may be used to compare the removals, across different treatment processes, of the related NOM fractions. Therefore, these ratios were computed and attempts were made to find out whether they are consistent with what is known about the removal (by different processes) of specific NOM fractions to which some of these fluorescent components have been ascribed. Paired t-tests were performed to ascertain whether there were statistically significant changes in the ratios of F_{max} across coagulation, ozonation and BAC filtration processes. Figure 4.5 shows the variation of ratios of F_{max} across the treatment train of two dominant humic-like components (C1 and C2) to that of protein-like components (C4 and C7), as well as of component C1 to that of C2.

During coagulation, the ratios of F_{max} of humic-like to that of protein-like components decreased for all cases except for humic-like component C3, which did not show a significant change relative to either of the protein-like components C4 or C7. The preferential reduction of humic-like components is consistent with the preferential removal of hydrophobic high molecular weight humic NOM by coagulation (Allpike et al., 2005; Bolto et al., 2002). Components C1 and C2 were preferentially removed relative to all the other humic-like components except in one case: there was no significant difference between the reduction of components C2 and C6. This may be an indication that C1 and C2 are representative of larger molecular weight and more humic compounds, which have been found to be preferentially removed by coagulation (Haberkamp et al., 2007 ; Humbert et al., 2007).

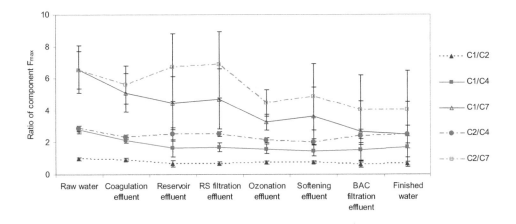

Figure 4.27 Variation of ratios of F_{max} across the treatment train of the two dominant humic-like components (C1 and C2) to that of protein-like components (C4 and C7), as well as of component C1 to that of C2.

Ozonation degraded humic-like components C1, C2, and C6 more than protein-like components. While degradation of C5 was less, that of C3 did not differ significantly from that of protein-like components. There was no significant difference between the rates of ozone-degradation of protein-like components C4 and C7. The preferential reduction of humic fluorescence is consistent with the lower reactivity of microbially derived NOM (represented by protein-like fluorophores) with ozone (Bose and Reckhow, 2007).

During BAC filtration, the ratios between F_{max} of humic-like components did not show significant changes except for two cases: the ratio of F_{max} of C1 to that of C3 decreased ($p <$ 0.01), while the ratio of F_{max} of C2 to that of C6 increased ($p < 0.05$). The ratios of humic-like to protein-like components did not change significantly except for components C1 and C5 ($p < 0.05$ and $p < 0.01$, respectively), which decreased relative to the tyrosine-like component C7. There was significant ($p < 0.05$) reduction in fluorescence intensity of the tryptophan-like component C4 relative to that of tyrosine-like component C7. These results appear to indicate that humic-like components were removed by BAC filtration just as effectively as protein-like components. Because the BAC filters at Weesperkarspel water treatment plant are operated for extended periods before regeneration (more than six months), NOM is considered to be removed mainly by biodegradation, although adsorption may also play a role. It would therefore be expected that, because it is generally not easily biodegradable, aromatic humic NOM (represented by humic-like components) would not be as well removed as microbially derived NOM (represented by protein-like components). That this is not apparent from the results could be due to one or a combination of factors: release of fluorescent bacterial exudates from the biofilter could offset the preferential reduction of humic-like fluorescence; presence of considerable variances in the analytical measurements of protein-like fluorescence; and persistence of protein-like fluorescence signature, which has been used as a tracer of microbial organic matter from wastewater pollution. In a comparative study of removal of effluent organic matter from tertiary effluent of a wastewater treatment plant by direct nanofiltration (NF) and powdered activated carbon/NF, the signature of wastewater was detectable as protein-like fluorescence even at a very low DOC concentration of < 0.5 mg C/L in the permeate (Kazner et al., 2008).

4.3.4 Correlations

All samples from the pre-treatment and post treatment plants were included in the determination of Spearman's correlation coefficients. Table 4.3 is a correlation matrix obtained with SPSS statistical software. There were significant correlations (p < 0.01) among sample DOC concentration, UVA_{254}, and F_{max} for the seven PARAFAC components (C1, C2, C3, C4, C5, C6, and C7) and DOC concentrations for the five SEC-OCD fractions (humics, building blocks, neutrals, biopolymers and LMW acids).

Table 4.3 Correlation matrix of DOC, UVA_{254}, and F_{max} with the seven PARAFAC components and concentrations of SEC-OCD fractions for all samples from the water treatment train.

	DOC	UVA_{254}	C1	C2	C3	C4	C5	C6	C7	Humics	Building blocks	Neutrals	Biopolymers	LMW acids
DOC	1.00	.88*	.96*	.97*	.94*	.94*	.95*	.93*	.91*	.98*	.82*	.96*	.78*	-.68*
UVA_{254}		1.00	.89*	.91*	.88*	.92*	.89*	.91*	.86*	.85*	.67*	.84*	.78*	-.69*
C1			1.00	.97*	.98*	.95*	.97*	.96*	.89*	.93*	.77*	.95*	.76*	-.72*
C2				1.00	.95*	.97*	.97*	.94*	.93*	.96*	.78*	.92*	.76*	-.72*
C3					1.00	.96*	.97*	.96*	.86*	.91*	.78*	.92*	.76*	-.71*
C4						1.00	.96*	.95*	.93*	.90*	.76*	.91*	.81*	-.73*
C5							1.00	.96*	.89*	.93*	.78*	.92*	.72*	-.71*
C6								1.00	.86*	.90*	.74*	.92*	.76*	-.72*
C7									1.00	.89*	.72*	.86*	.76*	-.68*
Humics										1.00	.77*	.92*	.69*	-.65*
Building blocks											1.00	.82*	.66*	-.40*
Neutrals												1.00	.81*	-.65*
Biopolymers													1.00	-.53*
LMW acids														1.00

* Correlation is significant at the 0.01 level (2-tailed).

The correlations of humics, building blocks and neutral fractions were higher with DOC than with UVA_{254} or with F_{max} for any of the PARAFAC components. This result would be expected since measurements of DOC and SEC-OCD fractions are all based on detection of carbon dioxide (CO_2) produced by photo-oxidation of organic carbon, while UVA_{254} and F_{max} measure only a part of organic matter responsible for UV absorption and fluorescence, respectively. The biopolymer fraction did not display a similar trend but, rather, correlated more or less equally with DOC, UVA_{254} and F_{max}.

DOC correlated nearly perfectly ($r = 0.98$, $p < 0.01$) with humic fraction but not as highly ($r = 0.78$, $p < 0.01$) with biopolymer fraction; the latter displayed more variability for pre-treatment plant water samples (DOC > 6.0 mg C/L) (Figure 4.6). This difference in degree of correlation with DOC between humic and biopolymer fractions could be due to a lower oxidation efficiency of the latter in the DOC detector of the SEC-OCD system, which uses UV oxidation to decompose organic carbon to CO_2 which is then measured by nondispersive infrared absorption. In a study to evaluate the performance of an online DOC detector for detection of NOM samples using a similar SEC-OCD system, the highest molecular weight biopolymer fraction (attributed mainly to polysaccharides) was found to be poorly oxidized, thus underestimating its concentration on the basis of the detected DOC (Lankes et al., 2009).

DOC correlated slightly higher with F_{max} than with UVA_{254}. DOC, UVA_{254} and F_{max} correlated more strongly with humics and neutrals than with building blocks and biopolymers. The terrestrial humic-like component C2 (as well as C1 and C3) showed slightly better predictions of DOC and humic fraction concentrations than did UVA_{254} (Figure 4.7). The latter showed more variability for pre-treatment plant samples (DOC > 6.0 mg C/L and humics > 4.0 mg C/L).

The tryptophan-like and tyrosine-like components C4 and C7 correlated positively with biopolymer fraction (Figure 4.8). Since tryptophan-like and tyrosine-like fluorescence have been found to correlate with protein-like NOM, this might be an indication of some input of microbial NOM in the samples analyzed. In an evaluation of pyrolysis gas chromatography mass spectrometry (Py/GC/MS) products of soil-water and streamwater NOM, it was found that PARAFAC protein-like components correlated significantly ($p<0.05$) with nitrogen containing compounds but PARAFAC components correlated poorly with polysaccharide content (Fellman et al., 2009). However, in spite of the strong correlation of the protein-like fluorescence with biopolymer concentration, there was still a high percentage of variation in biopolymer measurements which could be partly attributed to the presence of non-fluorescing polysaccharides in the biopolymer fraction.

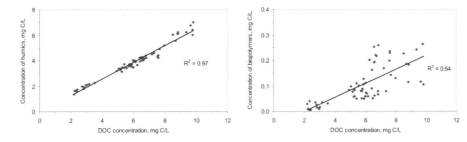

Figure 4.28 Regressions describing the relationship between DOC concentrations and concentrations of humics and biopolymers.

The higher predictive power of F_{max} provides an opportunity for its use as an alternative to UVA_{254} as a surrogate measure of DOC for online monitoring of its concentration in drinking water treatment plants. Furthermore, the higher sensitivity of fluorescence measurements allows measurements of very low NOM concentrations. The correlation of F_{max} of protein-like components with the biopolymer fraction, which may include nitrogen containing compounds, further demonstrates its potential for online monitoring of sub-fractions of DOC which are known to be more labile, thus promoting biogrowth in distribution systems, and to contribute to irreversible protein fouling of polymeric water filtration membranes. Whereas this study used offline measurements to generate F-EEMs, it is possible to develop online methods for near real time monitoring, thus allowing operational changes to be made whenever required.

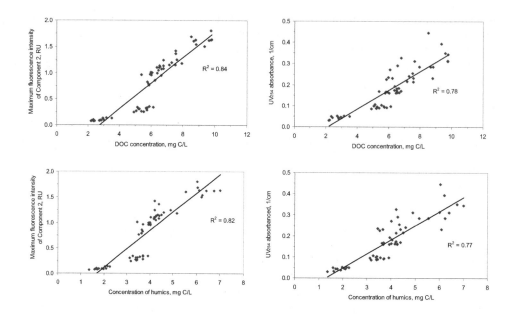

Figure 4.29 Regressions describing the relationships between (top) DOC, F_{max} of PARAFAC component C2 and UVA$_{254}$ absorbance, and (bottom) humic fraction concentration, F_{max} of PARAFAC component C2 and UVA$_{254}$ absorbance.

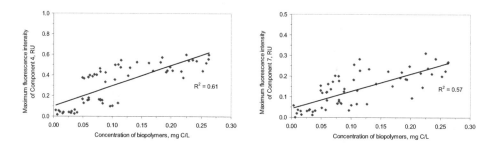

Figure 4.30 Regressions describing the relationship between concentrations of biopolymer fraction and (left) F_{max} of tryptophan-like component C4 and (right) F_{max} of tyrosine-like component C7.

4.4   Conclusions

Based on the characterization of NOM in water samples from a drinking water treatment plant using F-EEMs and PARAFAC and investigation of the correlation between extracted fluorescent component and NOM fractions obtained using SEC-OCD, the following conclusions can be drawn from this study:

- F-EEMs, of samples from Loenderveen/Weesperkarspel drinking water treatment plants, and PARAFAC were used to develop a 7-component model in which the components have fluorescence spectra similar to those of fluorescent components extracted in previous studies.

- 5 of the components are humic-like and two are protein-like (tryptophan-like and tyrosine-like).

- There were significant correlations ($p < 0.01$) between sample DOC concentration, UVA_{254}, and F_{max} for the seven PARAFAC components and DOC concentrations of the five SEC-OCD fractions.

- Three of the humic-like components showed slightly better predictions of DOC and of humic fraction concentrations than UVA_{254}.

- Tryptophan-like and tyrosine-like components correlated positively ($r = 0.78$ and 0.75, respectively) with biopolymer fraction.

- Except for component C3, which did not change significantly relative to protein-like components, there was preferential reduction of humic-like relative to protein-like components during coagulation.

- There was a reduction in the ratio of F_{max} of all humic-like, except C3 and C5, relative to that of protein-like components by ozonation, indicating, in general, stronger reactivity of ozone with humic-like NOM.

- During BAC filtration, the ratios of F_{max} of humic-like to protein-like components did not change significantly except for components C1 and C5, which decreased relative to tyrosine-like component C7.

- There is need for further research on the identities of fluorescent components obtained with PARAFAC and how they relate with NOM characteristics determined using alternative techniques.

4.5   References

Abbt-Braun, G., Lankes, U. and Frimmel, F.H. 2004 Structural characterization of aquatic humic substances – The need for a multiple method approach. *Aquatic Sciences* 66, 151-170.

Allpike, B.P., Heitz, A., Joll, C.A., Kagi, R.I., Abbt-Braun, G., Frimmel, F.H., Brinkmann, T., Her, N. and Amy, G. 2005 Size Exclusion Chromatography To Characterize DOC Removal in Drinking Water Treatment. *Environ. Sci. Technol.* 39(7), 2334-2342.

Andersen, C.M. and Bro, R. 2003 Practical aspects of PARAFAC modeling of fluorescence excitation-emission data. *Journal of Chemometrics* 17(4), 200-215.

Baker, A. 2001 Fluorescence excitation-emission matrix characterization of some sewage-impacted rivers. *Environ. Sci. Technol.* 35(5), 948-953.

Bieroza, M., Baker, A. and Bridgeman, J. 2009a Relating freshwater organic matter fluorescence to organic carbon removal efficiency in drinking water treatment. *Sci. Total Environ.* 407(5), 1765-1774.

Bieroza, M., Baker, A. and Bridgeman, J. 2009b Exploratory analysis of excitation-emission matrix fluorescence spectra with self-organizing maps as a basis for determination of organic matter removal efficiency at water treatment works. *J. Geophys. Res.,* 114.

Bieroza, M.Z., Bridgeman, J. and Baker, A. 2010 Fluorescence spectroscopy as a tool for determination of organic matter removal efficiency at water treatment works. *Drink. Water Eng. Sci.* 3, 63-70.

Bolto, B., Dixon, D., Eldridge, R. and King, S. 2002 Removal of THM precursors by coagulation or ion exchange. *Water Res.* 36(20), 5066-5073.

Borisover, M., Laor, Y., Parparov, A., Bukhanovsky, N. and Lado, M. 2009 Spatial and seasonal patterns of fluorescent organic matter inLake Kinneret (Sea of Galilee) and its catchment basin. *Water Res.* 43, 3104-3116.

Bose, P. and Reckhow, D.A. 2007 The effect of ozonation on natural organic matter removal by alum coagulation. *Water Res.* 41, 1516 – 1524.

Bro, R. 1997 PARAFAC. Tutorial and applications. *Chemometrics and Intelligent Laboratory Systems* 38(2), 149-171.

Chen, W., Westerhoff, P., Leenheer, J.A. and Booksh, K. 2003 Fluorescence Excitation-Emission Matrix Regional Integration to Quantify Spectra for Dissolved Organic Matter. *Environ. Sci. Technol.* 37, 5701-5710.

Chin, Y.-P., Aiken, G. and O'Loughlin, E. 1994 Molecular Weight, Polydispersity, and Spectroscopic Properties of Aquatic Humic Substances. *Environ. Sci. Technol.* 28, 1853-1858.

Coble, P.G., Green, S.A., Blough, N.V. and Gagosian, R.B. 1990 Characterization of dissolved organic matter in the Black Sea by fluorescence spectroscopy. *Nature* 348, 432-435.

Coble, P.G. 1996 Characterization of marine and terrestrial DOM in seawater using excitation-emission matrix spectroscopy. *Marine Chemistry* 51, 325-346.

Cory, R.M. and McKnight, D.M. 2005 Fluorescence Spectroscopy Reveals Ubiquitous Presence of Oxidized and Reduced Quinones in Dissolved Organic Matter. *Environ. Sci. Technol.* 39, 8142-8149.

Croué, J.-P., G.V.Korshin and M.M.Benjamin (eds) (2000) Characterization of Natural Organic Matter in Drinking Water, AwwRF, Denver, CO.

Croué, J.-P. 2004 Isolation of humic and non-humic NOM fractions: Structural characterization. *Environmental Monitoring and Assessment* 92(1-3), 193-207.

Fellman, J.B., Miller, M.P., Cory, R.M., D'Amore, D.V. and White, D. 2009 Characterizing Dissolved Organic Matter Using PARAFAC Modeling of Fluorescence Spectroscopy: A Comparison of Two Models. *Environ. Sci. Technol.* 43(16), 6228-6234.

Frimmel, F.H. 1998 Characterization of natural organic matter as major constituents in aquatic systems. *Journal of Contaminant Hydrology* 35, 201–216.

Frimmel, F.H. and Abbt-Braun, G. 1999 Basic Characterization of Reference NOM from Central Europe - Similarities and Differences. *Environment International* 25(2/3), 191-207.

Haberkamp, J., Ruhl, A.S., Ernst, M. and Jekel, M. 2007 Impact of coagulation and adsorption on DOC fractions of secondary effluent and resulting fouling behaviour in ultrafiltration. *Water Res.* 41, 3794-3802.

(1984) "How can I know if it's real?" A catalog of diagnostics for use with three-mode factor analysis and multidimensional scaling, . In H. G. Law, C. W. Snyder, J. A. Hattie, and R. P. McDonald (eds.), Research methods for multimode data analysis. Praeger, New York, pp. 566-591.

Henderson, R.K., Baker, A., Murphy, K.R., Hambly, A., Stuetz, R.M. and Khan, S.J. 2009 Fluorescence as a potential monitoring tool for recycled water systems: A review. *Water Res.* 43 863-881.

Her, N., Amy, G., McKnight, D., Sohna, J. and Yoon, Y. 2003 Characterization of DOM as a function of MW by fluorescence EEM and HPLC-SEC using UVA, DOC, and fluorescence detection. *Water Res.* 37, 4295–4303.

Huber, S.A. and Frimmel, F.H. 1994 Direct Gel Chromatographic Characterization and Quantification of Marine Dissolved Organic Carbon Using High-Sensitivity DOC Detection. *Environ. Sci. Technol.* 28(6), 1194-1197.

Hudson, N., Baker, A. and Reynolds, D. 2007 Fluorescence analysis of dissolved organic matter in natural, waste and polluted waters - A review. *River Research and Applications* 23(6), 631-649.

Humbert, H., Gallard, H., Jacquemet, V.r. and Croue, J.-P. 2007 Combination of coagulation and ion exchange for the reduction of UF fouling properties of a high DOC content surface water. *Water Res.* 41, 3803 – 3811.

Hunt, J.F. and Ohno, T. 2007 Characterization of fresh and decomposed dissolved organic matter using excitation-emission matrix fluorescence spectroscopy and multiway analysis. *J. Agricultural and Food Chemistry* 55(6), 2121-2128.

Kazner, C., Baghoth, S., Sharma, S., Amy, G., Wintgens, T. and Melin, T. 2008 Comparing the effluent organic matter removal of direct NF and powdered activated carbon/NF as high quality pretreatment options for artificial groundwater recharge. *Water Sci. Technol. Water Supply* 57(6), 821-827.

Lankes, U., Mueller, M.B., Weber, M. and Frimmel, F.H. 2009 Reconsidering the quantitative analysis of organic carbon concentrations in size exclusion chromatography. *Water Res.* 43(4), 915-924.

Leenheer, J.A. 2004 Comprehensive assessment of precursors, diagenesis, and reactivity to water treatment of dissolved and colloidal organic matter. *Water Sci. Technol. Water Supply* 4(4), 1-9.

Lorenzo-Seva, U. and Ten Berge, J.M.F. 2006 Tucker's congruence coefficient as a meaningful index of factor similarity. *Methodology* 2, 57-64.

Matilainen, A., Lindqvist, N., Korhonen, S. and Tuhkanen, T. 2002 Removal of NOM in the different stages of the water treatment process. *Environment International* 28, 457– 465.

McKnight, D.M., Boyer, E.W., Westerhoff, P.K., Doran, P.T., Kulbe, T., Andersen, D.T. and Andersen, D.T. 2001 Spectrofluorometric characterization of dissolved organic matter for indication of precursor organic material and aromaticity. *Limnol. Oceanogr.* 46(1), 38-48.

Murphy, K.R., Ruiz, G.M., Dunsmuir, W.T.M. and Waite, T.D. 2006 Optimized parameters for fluorescence-based verification of ballast water exchange by ships. *Environ. Sci. Technol.* 40(7), 2357-2362.

Murphy, K.R., Stedmon, C.A., Waite, T.D. and Ruiz, G.M. 2008 Distinguishing between terrestrial and autochthonous organic matter sources in marine environments using fluorescence spectroscopy. *Marine Chemistry* 108(1-2), 40-58.

Owen, D.M., Amy, G.L. and Chowdhary, Z.K. (eds) (1993) Characterization of Natural Organic Matter and its Relationship to Treatability, American Water Works Association Research Foundation, Denver, CO.

Peiris, R.H., Halle, C., Budman, H., Moresoli, C., Peldszus, S., Huck, P.M. and Legge, R.L. 2010 Identifying fouling events in a membrane-based drinking water treatment process using principal component analysis of fluorescence excitation-emission matrices. *Water Res.* 44, 185-194.

Persson, T. and Wedborg, M. 2001 Multivariate evaluation of the fluorescence of aquatic organic matter. *Analytica Chimica Acta* 434(2), 179-192.

Peuravuori, J., Koivikko, R. and Pihlaja, K. 2002 Characterization, differentiation and classification of aquatic humic matter separated with different sorbents: synchronous scanning fluorescence spectroscopy. *Water Res.* 36, 4552–4562.

Stedmon, C.A., Markager, S. and Bro, R. 2003 Tracing dissolved organic matter in aquatic environments using a new approach to fluorescence spectroscopy. *Marine Chemistry* 82, 239–254.

Stedmon, C.A. and Markager, S. 2005a Resolving the variability in dissolved organic matter fluorescence in a temperate estuary and its catchment using PARAFAC analysis. *Limnol. Oceanogr.* 50(2), 686-697.

Stedmon, C.A. and Markager, S. 2005b Tracing the production and degradation of autochthonous fractions of dissolved organic matter by fluorescence analysis. *Limnol. Oceanogr.* 50(5), 1415–1426.

Stedmon, C.A., Thomas, D.N., Granskog, M., Kaartokallio, H., Papadimitriou, S. and Kuosa, H. 2007 Characteristics of dissolved organic matter in Baltic coastal sea ice: Allochthonous or autochthonous origins? *Environ. Sci. Technol.* 41(21), 7273-7279.

Stedmon, C.A. and Bro, R. 2008 Characterizing dissolved organic matter fluorescence with parallel factor analysis: a tutorial. *Limnol. Oceanogr.: Methods* 6, 572-579.

Swietlik, J., Dabrowska, A., Raczyk-Stanislawiak, U. and Nawrocki, J. 2004 Reactivity of natural organic matter fractions with chlorine dioxide and ozone. *Water Res.* 38(3), 547-558.

Vuorio, E., Vahala, R., Rintala, J. and Laukkanen, R. 1998 The evaluation of drinking water treatment performed with HPSEC. *Environment International* 24(5/6), 617-623.

Wu, F.C., Evans, R.D. and Dillon, P.J. 2003 Separation and Characterization of NOM by High-Performance Liquid Chromatography and On-Line Three-Dimensional Excitation Emission Matrix Fluorescence Detection. *Environ. Sci. Technol.* 37, 3687-3693.

Yamashita, Y. and Tanoue, E. 2003 Chemical characterization of protein-like fluorophores in DOM in relation to aromatic amino acids. *Marine Chemistry* 82, 255-271.

Yamashita, Y. and Jaffe, R. 2008 Characterizing the Interactions Between Metals and Dissolved Organic Matter using Excitation#Emission Matrix and Parallel Factor Analysis. *Environ. Sci. Technol.* 42, 7374-7379.

Chapter 5

CHARACTERIZING NATURAL ORGANIC MATTER (NOM) AND REMOVAL TRENDS DURING DRINKING WATER TREATMENT

A part of this chapter has been published as:

Baghoth, S.A., Sharma, S.K., Guitard, M., Heim, V., Croue, J.P. and Amy, G.L. 2011 Removal of NOM-constituents as characterized by LC-OCD and F-EEM during drinking water treatment. *J. Water Supply. Res. Technol. AQUA* 60(7), 412-424.

Summary

Natural organic matter (NOM) is of concern in drinking water because it causes adverse aesthetic qualities such as taste, odour, and colour; impedes the performance of treatment processes; and decreases the effectiveness of oxidants and disinfectants while contributing to undesirable disinfectants by-products. The effective removal of NOM during drinking water treatment requires a good understanding of its character. Because of its heterogeneity, NOM characterization necessitates the use of multiple analytical techniques. In this study, NOM in water samples from two drinking water treatment trains was characterized using size exclusion chromatography with organic carbon detection (SEC-OCD), and fluorescence excitation–emission matrices (F-EEMs) with parallel factor analysis (PARAFAC). These characterization methods indicate that the raw and treated waters are dominated by humic substances. The results show that whereas the coagulation process for both plants may be optimized for the removal of bulk DOC, it is not likewise optimized for the removal of specific NOM fractions. A five component PARAFAC model was developed for the F-EEMs, three of which are humic-like, while two are protein-like. These PARAFAC components and the SEC-OCD fractions represented effective tools for the performance evaluation of the two water treatment plants in terms of the removal of NOM fractions.

5.1   Introduction

Naturally occurring aquatic organic matter (NOM) is a heterogeneous mixture of compounds found abundantly in natural waters. NOM originates from living and dead plants, animals and microorganisms, and from the degradation products of these sources (Chow et al., 1999). NOM significantly affects water treatment processes such as coagulation, oxidation, adsorption, and membrane filtration. It affects drinking water quality in a number of ways. It contributes to formation of potentially carcinogenic disinfection by-products (DBPs) (Sharp et al., 2004), promotes biological regrowth in the water distribution system and contributes to colour, tastes and odours. The extent to which NOM affects water treatment processes depends on its quantity and physicochemical characteristics. NOM that is rich in aromatic structures such as carboxylic and phenolic functional groups have been found to be highly reactive with chlorine, thus forming forming DBPs (Reckhow et al., 1990). These aromatic structures are commonly present as a significant percentage of humic substances, which typically form over 50% of NOM. Hydrophobic and large molecular humic substances are enriched with aromatic structures and are readily removed by conventional drinking water treatment consisting of flocculation, sedimentation and filtration. In contrast, less aromatic hydrophilic NOM is more difficult to remove and is a major contributor of easily biodegradable organic carbon, which promotes microbiological regrowth in the distribution system.

It is now widely accepted that the efficiency of drinking water treatment is greatly influenced by the amount and character of NOM present in water. Consequently, many water treatment utilities monitor NOM in their source waters in order to optimize treatment processes. Typically, this optimization has been obtained using bulk water quality parameters such as dissolved organic carbon (DOC) concentration and ultraviolet absorbance at a wavelength of 254 nm (UVA_{254}). Specific ultraviolet absorbance (SUVA), which is obtained by dividing the

UVA$_{254}$ by the DOC concentration, is another bulk parameter that has been used as a surrogate for NOM composition and reactivity (Weishaar, 2003). It has been found to be a good indicator for hydrophobic, aromatic and high MW NOM fractions such as humic and fulvic acids (Weishaar, 2003; Traina et al., 1990). However, the use of these bulk parameters has limitations. Many waters may contain NOM with similar DOC concentrations or UVA$_{254}$ absorptivities but with different characteristics such as molecular weight and reactivity, resulting in different removal efficiencies during treatment. A better understanding of its quantity as well as character is therefore required to improve the performance of treatment processes and to optimize the removal of NOM.

Because of its heterogeneity and complexity, it is not practical to characterize NOM in terms of all of its molecular constituents. As such, it is commonly characterized into groups of compounds with similar physicochemical properties. Molar mass or molecular weight (MW) is one of the main properties used in studies designed to improve our understanding of the physicochemical characteristics of NOM. It influences the adsorption, bioavailability as well as other water treatment processes for the removal of NOM. Low MW (LMW) NOM molecules tend to be more hydrophilic and thus more biodegradable, while higher MW NOM molecules tend to be more aromatic and more hydrophobic, and have higher affinity for adsorption. LMW NOM has been shown to decrease the effectiveness of water treatment by activated carbon filtration as it competes for adsorption sites with target compounds (Newcombe et al., 1997) and is more difficult to remove by coagulation (Chow et al., 1999). Furthermore, LMW NOM is also a major source of easily biodegradable organic matter, which is known to promote bacterial regrowth in drinking water distribution systems (Volk et al., 2000). Since MW or molecular size (MS) distribution is an important characteristic of NOM, a number of tools have been developed to characterize NOM in terms of the MW of its fractions. These tools include ultrafiltration, vapour pressure osmometry, field flow fractionation, ultracentrifugation, small-angle X-ray scattering, and high performance size exclusion chromatography (HPSEC) (Chin et al., 1994; Her et al., 2003; Frimmel, 1998; Croué, 2004).

HPSEC has been widely used to determine the MW distribution of NOM from a variety of aquatic environments. It is more attractive than other analytical techniques because of the minimal sample preparation, small sample volumes, and ease and speed of analysis (O'Loughlin and Chin, 2001). It separates molecules mainly as a function of molecular size or molecular weight, with larger molecules eluting from the chromatographic column earlier than smaller ones. It is traditionally coupled with a fixed wavelength (254 nm) UV detector in order to determine the concentration of DOC. However, because absorbance depends on the molecular structure of the absorbing NOM species, UV detection cannot be used to determine DOC concentrations of NOM fractions which do not absorb in the UV spectrum. UV detection works well for humic NOM, with relatively high SUVA, but less so for non-humic NOM, with relatively low SUVA. In order to detect any type of organic carbon species, irrespective of whether it absorbs UV light or not, HPSEC may coupled with a DOC detector. This enables determination of DOC concentration of NOM fractions with low UV absorptivity, such as proteins, or with no UV absorptivity, such as polysaccharides.

The use of HPSEC with multiple detectors, such as UV and DOC, provides a powerful analytical tool for characterizing NOM fractions from a variety of aquatic environments. It has been shown to be very effective in following changes in the NOM distribution along drinking water treatment trains; it can capture the removal of highly reactive NOM (i.e., humic type structures) (Fabris et al., 2008), show the shift from high MW to low MW structures after oxidation processes (i.e., more biodegradable NOM) (Vuorio et al., 1998), and reveal the preferential removal of low MW DOC by biological filters (Buchanan et al., 2008). HPSEC coupled with UV and DOC detectors (SEC-OCD) has been used to fractionate NOM into five fractions: biopolymers (such as polysaccharides, polypeptides, proteins and amino sugars); humic substances (fulvic and humic acids); building blocks (hydrolysates of humic substances); LMW humic substances and acids; and low molecular weight neutrals (such as alcohols, aldehydes, ketones and amino acids) (Huber and Frimmel, 1994).

Fluorescence is another property that is frequently used for NOM characterization. The relatively low expense and high sensitivity of fluorescence measurements, coupled with rapid data acquisition of water samples at low natural concentrations, have made fluorescence spectrophotometry using fluorescence excitation-emission matrices (F-EEM) attractive for NOM characterization of water samples. This characterization has typically involved the use of excitation-emission wavelength pairs to identify fluorophores based on the location of fluorescence peaks on F-EEM contour plots (Coble, 1996). These peaks have been used to distinguish between humic-like NOM, with longer emission wavelengths (> 350 nm), and protein-like NOM, with shorter emission wavelengths ($\leq$ 350 nm). Other methods include: fluorescence regional integration (FRI) (Chen et al. 2003); multivariate data analysis (e.g. Principal Component Analysis, PCA, and Partial Least Squares regression, PLS) (Persson and Wedborg, 2001); and multi-way data analysis using parallel factor analysis (PARAFAC) (Stedmon et al., 2003). PARAFAC has been used to decompose F-EEMs into individual components some of which have been attributed to protein-like or humic-like NOM (Hunt and Ohno, 2007, Stedmon et al., 2003, Stedmon and Markager, 2005a, Stedmon et al., 2007a, Stedmon et al., 2007b, Yamashita et al., 2008).

The primary objective of this study was to characterize NOM in water samples taken across two drinking water treatment plants serving the suburbs of Paris. This was carried out in order improve our understanding of the character of the NOM and its temporal variation in waters treated by the two plants. A secondary objective was to evaluate the performance of the treatment processes in terms of NOM removal. Samples were collected from the two treatment plants and analyzed using bulk water quality parameters as well as SEC-OCD and F-EEM with PARAFAC.

5.2 Materials and methods

5.2.1 Sampling

Water samples were collected from two drinking water treatment plants of Syndicat des Eaux d'Ile de France (SEDIF) which supply drinking water to the suburbs of the city of Paris, France. The samples were collected between March 2008 and September 2009. The two plants, Choisy-le-Roi (CR) and Neuilly-sur-Marne (NM), comprise conventional treatment coupled with biofiltration using ozonation followed by BAC filtration. Figure 5.1 shows the treatment process scheme and the sampling points for CR treatment plant. The following water samples were collected monthly from CR: (i) raw water; (ii) preozonated water; (iii) settled water; (iv) sand filtered water; (v) ozonated water; (vi) biological activated carbon (BAC) filtered water; and (vii) product (finished) water. Figure 5.2 shows the treatment process scheme and the sampling points for NM treatment plant, from which the following samples were collected quarterly: (i) raw water; (ii) settled water; (iii) sand filtered water; (iv) ozonated water; (v) BAC filtered water; and (vi) product (finished) water. On average, seven samples were collected from CR every month and six samples every three months from NM.

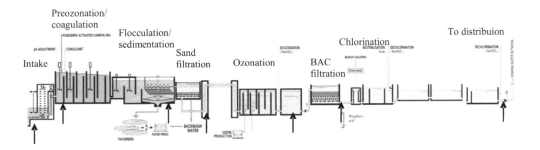

Figure 5.31 Treatment process scheme and sampling points for Choisy-le-Roi drinking water treatment plant.

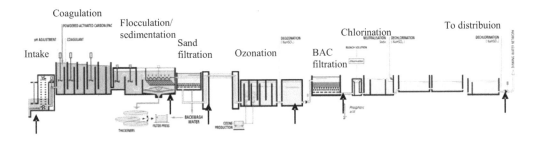

Figure 5.32 Treatment process scheme and sampling points for Neuilly-sur-Marne drinking water treatment plant.

The samples were collected in clean glass bottles and immediately filtered through 0.45 μm before being transported, within 24 hours, to the laboratory for analysis. The pre-filtered samples were stored at 5°C until required for analysis, which was normally done within one week of sampling. All the samples were analyzed for DOC concentration, UVA$_{254}$ and F-EEMs. Selected samples were analyzed using SEC-OCD. For CR, samples collected during 10 of the 18 months of sampling were analyzed (67 samples in total), while for NM, samples collected during 3 of the 7 quarters were analyzed (17 samples in total). Besides the data generated from these analyses, water quality data indicative of flooding (e.g., turbidity) and algal (e.g., chlorophyll a, cell counts) events for raw-water samples as well as routine parameters (calcium, alkalinity, conductivity, pH, temperature, ozone doses and coagulation doses) for raw and settled water samples were obtained from the two treatment plants.

5.2.2 DOC and UVA$_{254}$ measurements

DOC concentrations of all pre-filtered samples were determined by the catalytic combustion method using a Shimadzu TOC-V$_{CPN}$ organic carbon analyzer. UVA$_{254}$ of each sample was measured at room temperature (20±1°C) and ambient pH using a Shimadzu UV-2501PC UV-VIS scanning spectrophotometer. SUVA was determined by dividing the UVA$_{254}$ by the corresponding DOC concentration.

5.2.3 Characterization with SEC-OCD

Size exclusion chromatography of water samples was performed with a SEC-OCD system (DOC-LABOR, Germany) at Het Waterlaboratorium, Haarlem, The Netherlands. In the system, a small volume of the sample is injected in a TSK HW-50S chromatographic column which is connected to a Graentzel thin-film reactor (Huber and Frimmel, 1994). In the reactor, UV is used to oxidize NOM to CO_2, which is then measured by infrared detection, thus enabling the determination of DOC concentrations of the chromatographic fractions. The column separates NOM mainly according to molecular size/weight, and as many as five fractions may be fractionated. These fractions have been designated as: (i) biopolymers (BP), comprising polysaccharides, proteins and colloids, (ii) humic substances (HS), (iii) building blocks (hydrolysates of humics) (BB), (iv) low molecular weight humics and acids (LMW), and (v) low molecular weight neutrals (such as alcohols, aldehydes, ketones and amino acids). Besides the organic carbon detector, the system also incorporates a UV detector, which may be used to assess the aromaticity of the sample as well as of the humic fraction by computing the respective SUVA values, and a detector for measuring dissolved organic nitrogen (DON).

Water samples were analyzed without any pre-treatment other than filtration through 0.45 mm pore filters prior to injection in the chromatographic column. The classification of SEC-OCD fractions is based on empirical as well as systematic studies. For identification of HS fraction, up to five criteria may be used: (i) retention time, (ii) peak width, (iii) peak symmetry, (iv) the ratio of the peak area for the UV signal to that of the peak area for the DOC signal, and (v) DON. Definition of the fraction boundaries and quantification of the fractions by area integration of chromatograms was done with FIFFIKUS software (DOC-LABOR), which uses data for calibration standards as some of the inputs.

5.2.4 Fluorescence Excitation Emission Matrices (F-EEM)

Fluorescence intensities for all samples were measured at ambient pH and room temperature ($20\pm1^\circ$C) using a FluoroMax-3 spectrofluorometer (Horiba Jobin Yvon). To account for fluorescence quenching resulting from relatively high DOC concentration in water samples, absorbance corrections have to be applied to fluorescence measurements. However, these time-consuming corrections are not necessary if the sample UVA_{254} absorbance is less than 0.05 cm^{-1} (Kubista et al., 1994) or if the DOC concentration of the sample is diluted to about 1 mg C/L prior to fluorescence measurement (Westerhoff et al., 2001). Since UVA_{254} absorbance was more than 0.05 cm^{-1} for nearly all raw water samples from the two water treatment plants, the prefiltered samples were diluted to a DOC concentration of 1 mg C/L using ultrapure water obtained from a Milli-Q$^\circledR$ water purification system prior to fluorescence measurements.

F-EEMs were generated for each sample by scanning over excitation wavelengths between 240 and 450 nm at intervals of 10 nm and emission wavelengths between 290 and 500 nm at intervals of 2 nm. An F-EEM of Milli-Q$^\circledR$ water was obtained and this was subtracted from that of each sample in order to remove most of the water Raman scatter peaks. Since samples were diluted to a DOC concentration of 1 mg C/L prior to measurements, each blank subtracted F-EEM was multiplied by the respective dilution factor and Raman-normalized by dividing by the integrated area under the Raman scatter peak (excitation wavelength of 350 nm) of the corresponding Milli-Q$^\circledR$ water, and the fluorescence intensities reported in Raman units (RU).

5.2.5 PARAFAC modeling of fluorescence EEM

PARAFAC was used to model the dataset of F-EEMs generated for samples from both treatment plants. It uses an alternating least squares algorithm to minimize the sum of squared residuals in a trilinear model, thus allowing the estimation of the true underlying EEM spectra (Bro, 1997, Harshman and Lundy, 1994). It reduces a dataset of EEMs into a set of trilinear terms and a residual array (Andersen and Bro, 2003):

$$x_{ijk} = \sum_{f=1}^{F} a_{if}b_{jf}c_{kf} + \varepsilon_{ijk} \qquad i = 1,...,I; \ j = 1,...,J; \ k = 1,...,K$$

where x_{ijk} is the fluorescence intensity of the ith sample at the jth emission and kth excitation wavelength; a_{if} represents the concentration of the fth fluorophore in the ith sample (defined as scores), b_{jf} and c_{kf} are estimates of the emission and excitation spectra respectively for the fth fluorophore (defined as loadings), F is the number of fluorophores (components) and ε_{ijk} is the residual element, representing the unexplained variation in the model (Stedmon et al., 2003). While component scores indicate the relative concentrations of groups of organic fractions represented by the components, excitation and emission loadings indicate their characteristic excitation and emission spectra (Stedmon et al., 2003). The maximum fluorescence intensity for each component obtained from the PARAFAC analysis was used to illustrate the quantitative and qualitative differences between samples.

It is generally difficult to decide the most appropriate number of components of a PARAFAC model. There are several tools that may be used to select the appropriate number of components but only two were used in this study: the split-half analysis (Harshman, 1984), and the examination of residual error plots (Stedmon and Bro, 2008a). For split-half analysis, the dataset of EEMs was randomly split into two halves and a PARAFAC model obtained for each half. The excitation and emission spectral loadings of the two independent halves were then compared to ascertain whether they were similar.

A total of 179 water samples were collected for this study: 137 from CR and 42 from NM water treatment trains, respectively. A dataset comprising F-EEMs for 145 of these samples was used in the PARAFAC analysis. A series of PARAFAC models consisting of three to seven components were generated using DOMfluor toolbox (Stedmon and Bro, 2008b), which was specifically developed to perform PARAFAC analysis of DOM fluorescence. It contains all the tools used to perform split-half and residual errors diagnostics.

5.3   Results and discussion

5.3.1 Variation of DOC and SUVA

The mean DOC concentrations and SUVA values of the samples collected from CR and NM treatment process trains are shown in Table 5.1. Figure 5.3 shows the variation of DOC and SUVA for raw and product water samples collected from CR and NM treatment plants. In the case of CR, the DOC concentrations ranged between 2.0 mg C/L and 4.0 mg C/L for raw water, and between 1.0 mg C/L and 2.2 mg C/L for product water. The SUVA values varied between 1.7 and 3.7 L/mg-m for raw water, and between 0.2 and 2.0 L/mg-m for product water. For NM, the DOC concentrations ranged between 2.1 mg C/L and 3.2 mg C/L for raw water, and between 1.1 mg C/L and 1.6 mg C/L for product water. The SUVA values varied between 1.7 and 3.1 L/mg-m for raw water, and between 0.7 and 1.3 L/mg-m for product water. For both cases, there was no clear seasonal pattern in the variation of either DOC concentration or SUVA for the raw water. The treated-water DOC concentrations were fairly stable, indicating that the two treatment plants were generally effective in maintaining a relatively constant product water DOC concentration. On the contrary, the SUVA values were not stable, indicating that the plants were not as effective in maintaining a more uniform NOM character in the treated water.

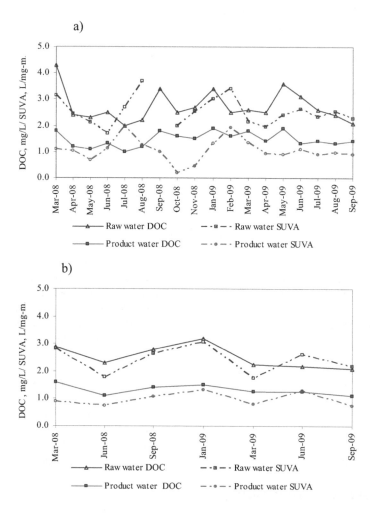

Figure 5.33 Variation of DOC concentrations and SUVA values for raw and treated water samples collected from (a) Choisy-le-Roi and (b) Neuilly-sur-Marne treatment plants.

Table 5.1 Mean value and standard deviations of DOC concentrations and SUVA values for water samples from Choisy-le-Roi and Neuilly-sur-Marne process trains.

Water Sample	Choisy-le-Roi, (n=20)		Neuilly-sur-Marne, (n=7)	
	DOC (mg C/L)	SUVA (L/mg-m)	DOC (mg C/L)	SUVA (L/mg-m)
Raw water	2.7 ± 0.6	2.5 ± 0.5	2.5 ± 0.4	2.4 ± 0.5
Preozonated water	2.8 ± 0.6	1.9 ± 0.3		
Settled water	2.1 ± 0.3	1.5 ± 0.4	1.9 ± 0.2	1.8 ± 0.3
Sand filtered water	1.9 ± 0.4	1.6 ± 0.5	1.6 ± 0.3	1.9 ± 0.4
Ozonated water	1.8 ± 0.3	1.2 ± 0.6	1.6 ± 0.2	1.1 ± 0.3
BAC filtered water	1.5 ± 0.3	1.2 ± 0.5	1.3 ± 0.2	1.1 ± 0.3
Product (finished) water	1.6 ± 0.3	1.1 ± 0.4	1.3 ± 0.2	1.0 ± 0.3

5.3.2 Treatment efficiencies in terms of DOC removal and SUVA reduction

For both treatment plants, the operational objective for the removal of NOM is to maintain a TOC concentration of $\leq 2.0 \pm 0.2$ mg C/L, a French drinking water treatment goal, as long as the raw water TOC concentration is ≤ 5.0 mg C/L. Since the TOC and DOC concentrations did not differ by more than 5% for any of the samples analyzed, the treatment efficiencies for the two process trains were assessed in terms of DOC removal and SUVA reduction.

The mean DOC and the mean percentage DOC removals across each treatment process for both CR and NM process trains are shown in Figure 5.4. In both plants, DOC was removed mainly by the coagulation-filtration process, which removed on average 0.9 mg C/L in CR and 0.8 mg C/L in NM, and by BAC filtration, which removed on average a further 0.3 mg C/L in CR and 0.2 mg C/L in NM. The results show that whereas the mean DOC concentrations of the raw and product waters for CR were slightly higher than for NM, the difference in the DOC removal efficiencies for the two treatment trains were not statistically significant for any of the processes. The maximum DOC concentrations of the raw and treated waters were 4.0 mg C/L and 2.2 mg C/L, respectively, for CR, and 3.2 mg C/L and 2.1 mg C/L, respectively, for NM. While the latter satisfies the treatment objective stated earlier, the fact that it was obtained when the raw water DOC concentration was only 3.1 mg C/L implies that it is likely that the objective could be compromised if the raw water DOC concentration approached 5 mg C/L, the maximum for which the treatment objective was optimized. Since the maximum DOC concentrations of the raw and treated waters did not coincide, the character of the NOM likely influenced the removal efficiencies.

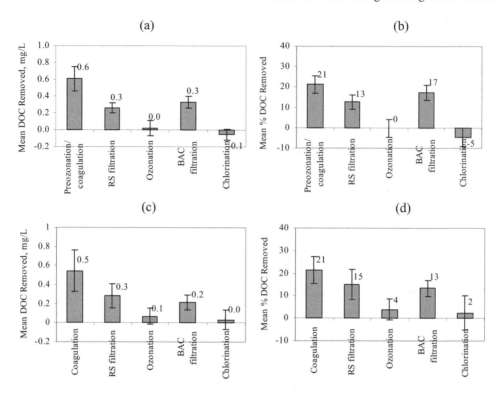

Figure 5.34 Mean DOC and mean percentage DOC removal at each water treatment step for water samples collected from Choisy-le-Roi ((a) and (b), respectively) and Neuilly-sur-Marne ((c) and (d), respectively) treatment trains. The error bars indicate 95% confidence interval.

SUVA is one of the main NOM characteristics that give an indication of its treatability by coagulation. It has been found to correlate with NOM aromaticity (Westerhoff et al., 1999) and is indicative of its hydrophobicity. Edzwald and Tobiason (1999) proposed some guidelines on the character of NOM and the expected removal of DOC by coagulation based on SUVA values. A SUVA value of about 4 L/mg-m or higher is indicative of NOM that is composed mainly of aquatic humics comprising large molecular weight highly hydrophobic humic acids. This NOM controls coagulation and has good DOC removal (>50% by alum coagulation). A value of 2-4 L/mg-m is composed of a mixture of aquatic and other NOM comprised of hydrophilic as well as hydrophobic fractions with mixed MW. This type of NOM influences coagulation and the DOC is fairly removed (25-50% using alum). A value of less than 2 L/mg-m comprises mainly of non-humic NOM with low hydrophobicity and low MW. This type of NOM has little influence on coagulation and is poorly removed (<25% as DOC by alum).

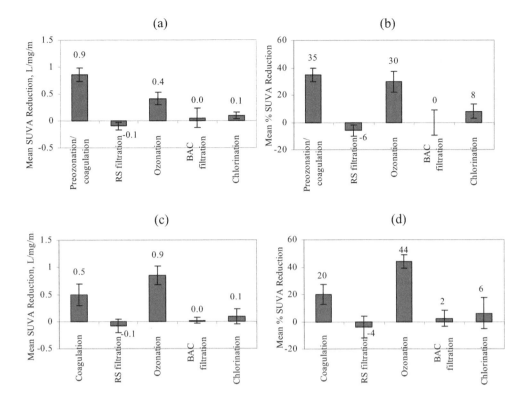

Figure 5.35 Mean SUVA reduction and mean percentage SUVA reduction at each water treatment step for water samples collected from Choisy-le-Roi ((a) and (b), respectively) and Neuilly-sur-Marne ((a) and (b), respectively) treatment trains. The error bars indicate 95% confidence interval.

The raw water SUVA ranged between 2 and 3.5 L/mg-m (Table 5.1) for both treatment plants, which would indicate that the NOM present in these waters was composed of hydrophilic as well as hydrophobic fractions. The mean SUVA reduction and the mean percentage SUVA reduction for samples collected from Choisy-le-Roi and Neuilly-sur-Marne are shown in Figure 5.5. The results show that, unlike DOC, which was removed mainly by coagulation-filtration and BAC filtration processes in both treatment plants (Figure 5.4), SUVA was reduced mainly by coagulation and by ozonation. BAC filtration did not have a significant effect on SUVA. The reduction of SUVA by coagulation is due to the preferential removal of larger MW hydrophobic NOM, while the reduction by ozonation is a consequence of the transformation of larger molecular weight fractions of NOM, which have higher UV absorptivities at 254 nm, to lower molecular weight fractions, which have lower UV absorptivities (Chandrakanth and Amy, 1996).

5.3.3 NOM characterization using SEC-OCD

SEC-OCD was used to obtain DOC concentrations of the five chromatographic fractions (biopolymers, humic substances, building blocks, LMW acids and LMW neutrals) before and after each treatment step for water samples from CR and NM. SEC-OCD analyses were performed for ten of the twenty and three of the seven months of sampling for CR and NM, respectively. Figures 5.6 and 5.7 show the mean DOC concentrations and the mean DOC removals for the SEC-OCD fractions across the treatment processes, and the percentage DOC contribution of the fractions in the raw and treated waters for the two process trains, respectively. Humic substances were the dominant fraction in all water samples from both plants, contributing on average to 55% of the DOC. Since the source waters for both plants are river water, it would be reasonable to expect the NOM composition to be typical of natural waters dominated by terrestrial runoff, in which humic substances (fulvic and humic acids) are 50% of the DOC (Thurman, 1985). In both plants, LMW acids were below the detection limit in nearly all samples.

The removal of the SEC-OCD fractions occurred mainly by coagulation, followed by BAC filtration. The trend in the change of NOM composition after treatment is similar for both process trains. The large MW fractions were preferentially removed, with the percentage contribution of the biopolymer fraction decreasing by a half, from 10 % to ~ 5% in the raw and treated waters, respectively, while that of humic substances decreased only slightly, by 1% for both trains. In contrast, there was a relative increase in the LMW fractions, with the building blocks increasing from 17% and 16% in the raw water for CR and NM, respectively, to 22% in the treated water for both plants, and the neutral fractions increasing slightly from 15% to 16% for both process trains. The reduction of the large MW hydrophobic humic substances and biopolymer fractions (which possibly include nitrogenous organic compounds) before chlorine disinfection, which is applied in both process trains, decreases the potential for formation of potentially carcinogenic disinfection by-products such as total trihalomethanes. The relative increase in the LMW fractions, which are generally more biodegradable, could potentially increase bacterial re-growth in the distribution system but the application of chlorination in both plants should minimize this.

To further evaluate the performance of the two process trains in terms of NOM removal, SEC-OCD data for a selection of sampling dates were examined in more detail. Since it removed the most DOC and is also a process that is routinely used to optimize DOC removal, the coagulation/flocculation process was used for the evaluation. In order to achieve the treatment objective of maintaining a TOC concentration of ≤ 2 mg/L in product water (there was no statistical difference between TOC and DOC for both plants), a calculated coagulant dose, which includes the raw water TOC as one of the parameters, is applied in both plants. As this objective was generally met on all the sampling dates for both plants, the performance was evaluated in terms of the removal efficiency of specific NOM fractions as measured by SEC-OCD.

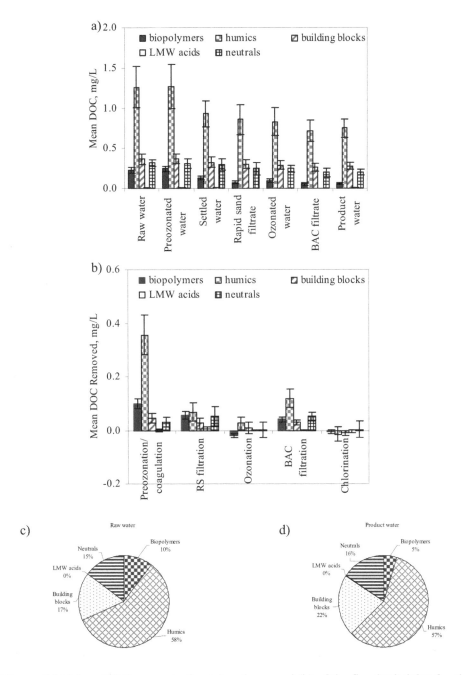

Figure 5.36 Mean DOC concentration (a) and removal (b) of the five SEC-OCD fractions for samples collected across Choisy-le-Roi treatment train, and the fractional DOC composition for raw (c) and treated (d) waters.

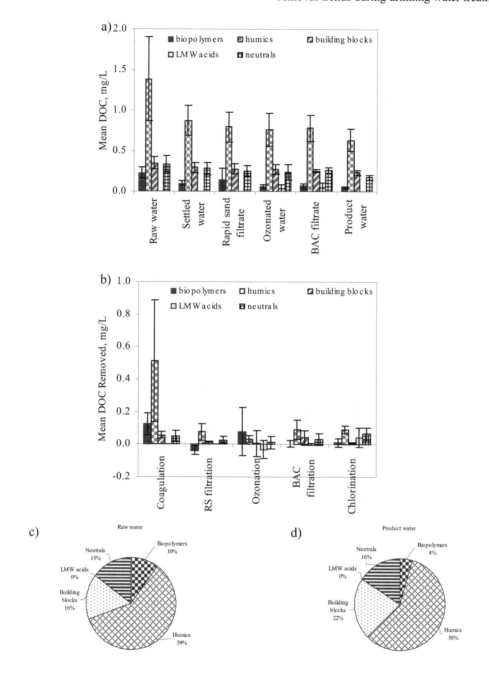

Figure 5.37 Mean DOC concentration (a) and removal (b) of the five SEC-OCD fractions for samples collected across Neuilly-sur-Marne treatment train, and the fractional DOC composition for raw (c) and treated (d) waters.

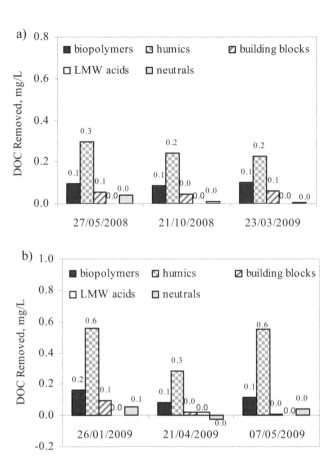

Figure 5.38 DOC of SEC-OCD fractions removed by coagulation process of Choisy-le-Roi treatment plant on three selected dates with (a) similar raw water DOC and coagulant dose, and (b) different raw water DOC and coagulant dose.

For the CR process train, the removal of SEC-OCD fractions by coagulation/flocculation was evaluated for two cases in each of which three samples were selected: (1) with similar raw water DOC concentrations and coagulant dosages (Figure 5.8a), and (2) with different raw water DOC concentrations and coagulant doses (Figure 5.8b). For the first case, the selected samples had DOC concentrations of 2.3-2.6 mg/L and the applied coagulant doses were 13.9-14.4 mg/L Al. The SUVA was ~ 2.0 L/(mg-m) and turbidity 2.4-4.8 NTU for all of the three samples. The similarity in DOC concentrations, the SUVA values, and the applied alum doses was reflected in the removal of the NOM fractions that are amenable to coagulation. The amounts of the large molecular weight biopolymer and humic fractions as well as the building blocks removed were similar for all of the three samples.

For the second case, comprising samples with different DOC concentrations and alum doses (Figure 5.8b), the DOC concentrations were ~3.5 mg/L for January and May 2009 samples, and 2.5 mg/L for April 2009 sample. The alum doses were ~ 32 mg/L Al for January and May 2009 samples, and 14.4 mg/L Al for April 2009 sample. The SUVA values were 3.0

L/(mg-m) for January, 2.0 L/(mg-m) for April, and 2.4 L/mg-m for May. The removal of humic substances (0.6 mg/L) was similar for January and May samples, indicating that the difference in SUVA values did not significantly affect the removal efficiency of this fraction. In contrast, whereas 0.1 mg/L of building blocks was removed for the January sample, hardly any was removed for the May sample, indicating that the lower SUVA for the latter may have made the removal of this fraction more difficult. The removal of humic substances for April sample (0.3 mg/L) was 50% of that for January or May, which is in roughly the same ratio as the applied alum doses. As for May sample, which had a similar SUVA, significantly less building blocks were removed for April as for January sample, which had a higher SUVA.

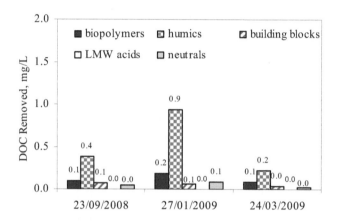

Figure 5.39 DOC of SEC-OCD fractions removed by coagulation process on three selected samples collected from Neuilly-sur-Marne water treatment plant.

For NM process train, SEC-OCD analyses were performed for three of the seven sets of samples and the DOC of SEC-OCD fractions removed by coagulation/flocculation are shown in Figure 5.9. For these sampling dates, the DOC concentrations and SUVA values were higher for January 2009 than for either September 2008 or March 2009 samples. Furthermore, the raw water turbidity was significantly higher for the January sample. Consequently, the alum dose for the sample of January (50 mg/L Al) was more than twice as much as for the other two samples (~20 mg/L Al). There was a correspondingly higher removal of the large molecular weight (biopolymers and humics) and neutral fractions for the January 2009 sample than for the other two samples dates. However, whereas the removal of biopolymers (normalized to the applied alum dose) was the same (0.004 mg C/L per mg/L Al) for the three samples, that of humics was the same (0.02 mg C/L per mg/L Al) for September 2008 and January 2009 but less (0.01 mg C/L per mg/L Al) for March 2009. The lower removal for the latter is consistent with its lower SUVA (1.7 L/(mg-m)) compared to that for September (2.6 L/(mg-m)) or January (3.1 L/(mg-m)).

5.3.4 Fluorescence EEMs

(a)   (b)

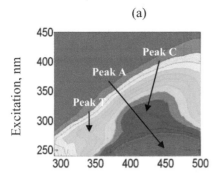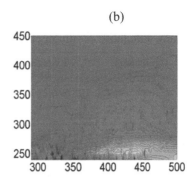

Emission wavelength, nm

Figure 5.40 Fluorescence EEM contour plots for raw (a) and treated (b) water samples collected from Choisy-le-Roi water treatment plant.

(a)   (b)

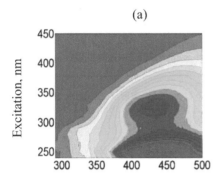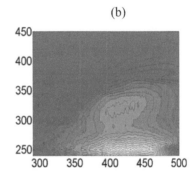

Emission wavelength, nm

Figure 5.41 Fluorescence EEM contour plots for raw (a) and treated (b) water samples collected from Neuilly-sur-Marne water treatment plant.

Three main fluorescence intensity peaks were obtained for all samples from both CR and Neuilly process trains that were analyzed. These previously identified peaks were observed at the following excitation and emission wavelengths: humic-like fluorescence (peak A) at 240-260 nm and 420-470 nm, respectively; fulvic-like fluorescence (peak C) at 300-340 nm and 400-450 nm respectively; and tryptophan-like fluorescence (peak T) at 240-280 nm and 300-360 nm, respectively. Figures 5.10 and 5.11 show typical contour plots of F-EEMs for raw and product water samples for CR and NM, respectively. In both cases, the fluorescence of the raw and treated waters was dominated by the humic-like peak A. There was substantial reduction of all of the three fluorescence peaks across the two treatment process trains. For both plants, the percentage reduction of the three peaks (relative to that of raw water) were similar across the treatment processes: 55% after coagulation/flocculation; 85% after BAC filtration; and 86% after chlorination (final water).

5.3.5 PARAFAC components extracted from fluorescence EEM

Table 5.2 Comparison of the spectral characteristics of five components identified in this study with those of similar components identified in previous studies.

Component of this study	Excitation/Emission wavelength (nm)	Description and source assignment (References)
Component 1	260(360)/480	Terrestrial humic substances
		Peak P3:<260(380)/498, (Ref. 3)
		Component 3: 270(360)/478, (Ref. 4)
		Component 3: 275(390)/479, (Ref. 7)
Component 2	250(320)/410	Terrestrial/anthropogenic humic substances
		Component 6: <250(320)/400, (Ref. 5)
		Component 2: 315/418, (Ref. 2)
		Component 3: 295/398, (Ref. 6)
		Component 3: 250(310)/400, (Ref. 9)
Component 3	<250(290)/360	Amino acids, free or protein bound
		Component 7: 240(300)/338, (Ref. 3)
		Component 4: <260(305)/378, (Ref. 8)
		Component 6: 250(290)/356, (Ref. 9)
Component 4	<250(300)/406	Terrestrial humic substances
		Component 1: <260(305)/428, (Ref. 8)
		Component 3:295/398, (Ref. 6)
		Peak C or M: (Ref. 1)
Component 5	270/306	Amino acids, free or protein bound
		Component 4: 275/306, (Ref. 6)
		Component 8: 275/304, (Ref. 5)
		Peak B: 275/310, (Ref. 1)

Secondary excitation wavelength is given in brackets.
Ref. 1. Coble, 1996, Ref. 2. Murphy et al., 2006, Ref. 3. Murphy et al., 2008, Ref. 4. Stedmon et al., 2003, Ref. 5. Stedmon and Markager, 2005a, Ref. 6. Stedmon and Markager, 2005b, Ref. 7. Yamashita et al., 2008, Ref. 8. Yamashita and Jaffe, 2008, Ref. 9. Kowalczuk et al., 2009.

A dataset comprising fluorescence EEMs for 145 water samples from both CR and NM were used for PARAFAC analysis. The analysis produced five models with the number of components in each ranging from 3 to 7. These models were subjected to a series of tests in order to determine the one with the most appropriate number of components. Split-half analyses were carried out for all the five models but only the three, the four and the five component models could be split-half validated. These were split-half validated in the sense that the corresponding components in the split halves had equal excitation and emission loadings as verified by the corresponding Tucker's congruence coefficients being greater than 0.95 (Lorenzo-Seva and Berge 2006). Of the three validated models, only the one with the highest (five) number of components was considered for further analysis.

The five components of the selected model have spectral features similar to those previously extracted from fluorescence EEMs of dissolved organic matter (DOM) (Borisover et al., 2009; Murphy et al., 2006; Murphy et al., 2008; Stedmon et al., 2003; Stedmon and Markager; 2005a, b; Stedmon et al., 2007b; Yamashita and Jaffe, 2008; Yamashita et al., 2008; Zhang et al., 2009). Table 5.2 shows a comparison of the excitation and emission wavelengths for the fluorescence maxima of the five components identified in this study with those of similar components identified in previous studies. Three humic-like components of terrestrial origin were identified: two dominant ones, component 1 (C1) and component 2 (C2); and a secondary one, component 4 (C4). Two of the components have excitation/emission characteristics similar to those of fluorescent protein-like compounds

(Cory and McKnight, 2005): component 3 (C3) is spectrally similar to tryptophan-like fluorophore; and component 5 (C5) is spectrally similar to tyrosine-like fluorophore.

The spectral contour plots and excitation and emission spectra of each of the identified components are shown in Figure 5.12. The spectra show relative fluorescence intensities (loadings), in Raman units, as a function of excitation and emission wavelengths for the complete dataset (solid) and for one of the independent halves used for validation (dotted).

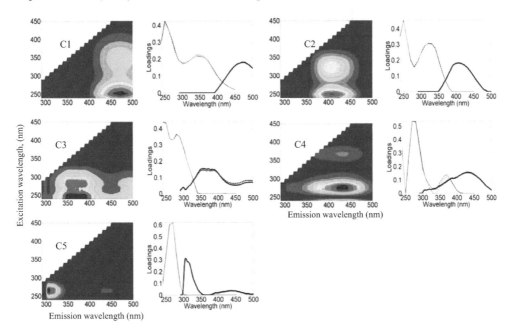

Figure 5.42 Contour plots of the five components identified by PARAFAC using the combined EEMs dataset of Choisy-le-Roi and Neuilly-sur-Marne samples. The line plots on the right show split-half validations of excitation (thin) and emission (thick) loadings between the complete dataset (solid) and one of the independent halves (dotted).

5.3.6 PARAFAC component scores across treatment

Figure 5.13 shows the average maximum fluorescence intensity (F_{max}) of the components across each process for CR and NM treatment trains. These fluorescence intensities give estimates of the relative concentrations of each component. For both plants, raw water samples exhibited higher F_{max} for terrestrial humic-like components C1 and C2 than for the other three components. Whereas the results appear to indicate that the raw water samples were dominated by humic-like fluorescent compounds, they are not sufficient to draw conclusions about the relative concentrations of all the seven components without prior knowledge of their respective quantum yields. However, results of SEC-OCD also showed quantitatively that for both treatment plants, humic substances comprised on average 50-60% of all samples analyzed. For both plants, F_{max} for tyrosine-like component C5 was generally stable across the treatment.

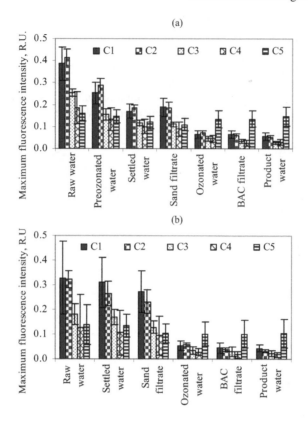

Figure 5.43 Maximum fluorescence intensities of PARAFAC components and their reductions during treatment at (a) Choisy-le-Roi and (b) Neuilly-sur-Marne treatment plants.

Figure 5.14 shows the mean percentage F_{max} reduction (relative to influent F_{max} at each process) for the five components across the two treatment plants. For CR, F_{max} for humic-like components C1, C2 and C4, and tryptophan-like component C3 were reduced by 30-60% after preozonation/coagulation, by 50-60% after ozonation, and by 10-30% after BAC filtration. It should be noted, however, that except for coagulation and filtration processes, the reduction in fluorescence does not always result in DOC reduction. Oxidation processes like ozonation only transform large molecular weight NOM into smaller and less aromatic organic compounds which have lower UV absorptivities and fluorescence. For NM, coagulation reduced F_{max} for components C1, C2, C3 and C4 by 15-30%, which is substantially less than that for CR. The higher reduction for the latter may be due mainly to the preozonation, which is applied in CR but not in NM; this may also partly explain why the reduction by ozonation is higher for Neuilly (~60-80%) than for CR. The effect of ozonation is not *intact* removal of a component but rather quenching of its fluorescence.

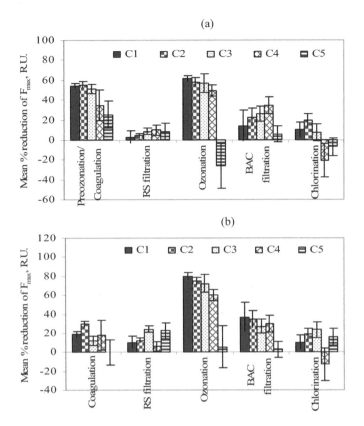

Figure 5.44 Mean percentage reduction in the maximum fluorescence intensities of PARAFAC components during treatment at (a) Choisy-le-Roi and (b) Neuilly-sur-Marne treatment plants.

5.4   Conclusions

Based on the characterization of NOM in water samples from CR and NM drinking water treatment plants using bulk NOM measurements, F-EEMs and SEC-OCD, the following conclusions can be drawn from this study:

- Whereas the treated water DOC concentrations were relatively stable for both treatment plants, indicating the effectiveness of bulk DOC removal, the SUVA values were not as stable, indicating that the NOM character of the treated water is more variable.

- Fluorescence and SEC-OCD measurements both showed that the raw water treated at the two water treatment plants is comprised mostly of humic substances.

- For both treatment plants, the large molecular weight fractions, comprising biopolymers and humic substances, were preferentially removed while the relative contribution of the low molecular weight fractions, comprising building blocks and neutrals, increased after treatment.

- SEC-OCD results indicate that for both plants, the coagulation process is not optimized for the removal of specific NOM fractions.

- A five component PARAFAC model of F-EEMs for samples from the CR and NM drinking water treatment plants was developed, comprising three humic-like and two protein-like substances (components)

- The fluorescence of samples from both treatment plants was dominated by terrestrial humic-like components, C1 and C2.

5.5 References

Andersen, C.M. and Bro, R. 2003 Practical aspects of PARAFAC modeling of fluorescence excitation-emission data. *Journal of Chemometrics* 17(4), 200-215.

Borisover, M., Laor, Y., Parparov, A., Bukhanovsky, N. and Lado, M. 2009 Spatial and seasonal patterns of fluorescent organic matter inLake Kinneret (Sea of Galilee) and its catchment basin. *Water Res.* 43, 3104-3116.

Bro, R. 1997 PARAFAC. Tutorial and applications. *Chemometrics and Intelligent Laboratory Systems* 38(2), 149-171.

Buchanan, W., Roddick, F. and Porter, N. 2008 Removal of VUV pre-treated natural organic matter by biologically activated carbon columns. *Water Res.* (42), 3335 – 3342.

Chandrakanth, M.S. and Amy, G.L. 1996 Effects of ozone on the colloidal stability and aggregation of particles coated with natural organic matter. *Environ. Sci. Technol.* 30(2), 431-443.

Chin, Y.-P., Aiken, G. and O'Loughlin, E. 1994 Molecular Weight, Polydispersity, and Spectroscopic Properties of Aquatic Humic Substances. *Environ. Sci. Technol.* 28, 1853-1858.

Chow, C.W.K., van Leeuwen, J.A., Drikas, M., Fabris, R., Spark, K.M. and Page, D.W. 1999 The impact of the character of natural organic matter in conventional treatment with alum. *Water Sci. Technol.* 40(9), 97-104.

Coble, P.G. 1996 Characterization of marine and terrestrial DOM in seawater using excitation-emission matrix spectroscopy. *Marine Chemistry* 51, 325-346.

Cory, R.M. and McKnight, D.M. 2005 Fluorescence Spectroscopy Reveals Ubiquitous Presence of Oxidized and Reduced Quinones in Dissolved Organic Matter. *Environ. Sci. Technol.* 39, 8142-8149.

Croué, J.-P. 2004 Isolation of humic and non-humic NOM fractions: Structural characterization. *Environmental Monitoring and Assessment* 92(1-3), 193-207.

Fabris, R., Chow, C.W.K., Drikas, M. and Eikebrokk, B. 2008 Comparison of NOM character in selected Australian and Norwegian drinking waters. *Water Res.* 42(15), 4188–4196.

Frimmel, F.H. 1998 Characterization of natural organic matter as major constituents in aquatic systems. *Journal of Contaminant Hydrology* 35, 201–216.

Harshman, R.A. (1984) *"How can I know if it's real?" A catalog of diagnostics for use with three-mode factor analysis and multidimensional scaling,* . In H. G. Law, C. W. Snyder, J. A. Hattie, and R. P. McDonald (eds.), Research methods for multimode data analysis. Praeger, New York, pp. 566-591.

Harshman, R.A. and Lundy, M.E. 1994 Parafac - Parallel Factor-Analysis. *Computational Statistics & Data Analysis* 18(1), 39-72.

Her, N., Amy, G., McKnight, D., Sohna, J. and Yoon, Y. 2003 Characterization of DOM as a function of MW by fluorescence EEM and HPLC-SEC using UVA, DOC, and fluorescence detection. *Water Res.* 37, 4295–4303.

Huber, S.A. and Frimmel, F.H. 1994 Direct Gel Chromatographic Characterization and Quantification of Marine Dissolved Organic Carbon Using High-Sensitivity DOC Detection. *Environ. Sci. Technol.* 28(6), 1194-1197.

Hunt, J.F. and Ohno, T. 2007 Characterization of fresh and decomposed dissolved organic matter using excitation-emission matrix fluorescence spectroscopy and multiway analysis. *J. Agricultural and Food Chemistry* 55(6), 2121-2128.

Kowalczuk, P., Durako, M.J., Young, H., Kahn, A.E., Cooper, W.J. and Gonsior, M. 2009 Characterization of dissolved organic matter fluorescence in the South Atlantic Bight with use of PARAFAC model: Interannual variability. *Marine Chemistry* 113(3-4), 182-196.

Kubista, M., Sjoback, R., Eriksson, S. and Albinsson, B. 1994 Experimental Correction for the Inner-Filter Effect in Fluorescence-Spectra. *Analyst* 119(3), 417-419.

Murphy, K.R., Ruiz, G.M., Dunsmuir, W.T.M. and Waite, T.D. 2006 Optimized parameters for fluorescence-based verification of ballast water exchange by ships. *Environ. Sci. Technol.* 40(7), 2357-2362.

Murphy, K.R., Stedmon, C.A., Waite, T.D. and Ruiz, G.M. 2008 Distinguishing between terrestrial and autochthonous organic matter sources in marine environments using fluorescence spectroscopy. *Marine Chemistry* 108(1-2), 40-58.

Newcombe, G., Drikas, M., Assemi, S. and Beckett, R. 1997 Influence of characterised natural organic material on activated carbon adsorption: I. Characterisation of concentrated reservoir water. *Water Res.* 31(5), 965-972.

O'Loughlin, E. and Chin, Y.P. 2001 Effect of detector wavelength on the determination of the molecular weight of humic substances by high-pressure size exclusion chromatography. *Water Res.* 35(1), 333-338.

Persson, T. and Wedborg, M. 2001 Multivariate evaluation of the fluorescence of aquatic organic matter. *Analytica Chimica Acta* 434(2), 179-192.

Reckhow, D.A., Singer, P.C. and Malcolm, R.L. 1990 Chlorination of Humic Materials - by-Product Formation and Chemical Interpretations. *Environ. Sci. Technol.* 24(11), 1655-1664.

Sharp, E.L., Parsons, S.A. and Jefferson, B. 2004 The effects of changing NOM composition and characteristics on coagulation performance, optimisation and control. *Water Science and Technology: Water Supply* 4(4), 95-102.

Stedmon, C.A., Markager, S. and Bro, R. 2003 Tracing dissolved organic matter in aquatic environments using a new approach to fluorescence spectroscopy. *Marine Chemistry* 82, 239–254.

Stedmon, C.A. and Markager, S. 2005a Resolving the variability in dissolved organic matter fluorescence in a temperate estuary and its catchment using PARAFAC analysis. *Limnol. Oceanogr.* 50(2), 686-697.

Stedmon, C.A. and Markager, S. 2005b Tracing the production and degradation of autochthonous fractions of dissolved organic matter by fluorescence analysis. *Limnol. Oceanogr.* 50(5), 1415–1426.

Stedmon, C.A., Markager, S., Tranvik, L., Kronberg, L., Slätis, T. and Martinsen, W. 2007a Photochemical production of ammonium and transformation of dissolved organic matter in the Baltic Sea. *Marine Chemistry* 104, 227–240.

Stedmon, C.A., Thomas, D.N., Granskog, M., Kaartokallio, H., Papadimitriou, S. and Kuosa, H. 2007b Characteristics of dissolved organic matter in Baltic coastal sea ice: Allochthonous or autochthonous origins? *Environ. Sci. Technol.* 41(21), 7273-7279.

Stedmon, C.A. and Bro, R. 2008a Characterizing dissolved organic matter fluorescence with parallel factor analysis: a tutorial. *Limnology and Oceanography-Methods* 6, 572-579.

Stedmon, C.A. and Bro, R. 2008b Characterizing dissolved organic matter fluorescence with parallel factor analysis: a tutorial. *Limnol. Oceanogr.: Methods* 6, 572-579.

Thurman, E.M. (1985) *Organic geochemistry of natural waters.* Martinus Nijhoff/Dr. W. Junk Publishers, Dordrecht (Netherlands).

Traina, S.J., Novak, J. and Smeck, N.E. 1990 An Ultraviolet Absorbance Method of Estimating the Percent Aromatic Carbon Content of Humic Acids. *J. Environ. Qual.* 19(1), 151-153.

Volk, C., Bell, K., Ibrahim, E., Verges, D., Amy, G. and Lechevallier, M. 2000 Impact of Enhanced and Optimized Coagulation on Removal of Organic Matter and its Biodegradable Fraction in Drinking Water. *Water Res.* 34(12), 3247-3257.

Vuorio, E., Vahala, R., Rintala, J. and Laukkanen, R. 1998 The evaluation of drinking water treatment performed with HPSEC. *Environment International* 24(5/6), 617-623.

Weishaar, J.L. 2003 Evaluation of Specific Ultraviolet Absorbance as an Indicator of the Chemical Composition and Reactivity of Dissolved Organic Carbon. *Environ. Sci. Technol.* 37, 4702-4708.

Westerhoff, P., Aiken, G., Amy, G. and Debroux, J. 1999 Relationships between the structure of natural organic matter and its reactivity towards molecular ozone and hydroxyl radicals. *Water Res.* 33(10), 2265-2276.

Westerhoff, P., Chen, W. and Esparza, M. 2001 Organic Compounds in the Environment Fluorescence Analysis of a Standard Fulvic Acid and Tertiary Treated Wastewater. *J. Environ. Qual.* 30, 2037–2046.

Yamashita, Y. and Jaffe, R. 2008 Characterizing the Interactions Between Metals and Dissolved Organic Matter using Excitation#Emission Matrix and Parallel Factor Analysis. *Environ. Sci. Technol.* 42, 7374-7379.

Yamashita, Y., Jaffe´, R., Maie, N. and Tanoue, E. 2008 Assessing the dynamics of dissolved organic matter (DOM) in coastal environments by excitation emission matrix fluorescence and parallel factor analysis (EEM-PARAFAC). *Limnol. Oceanogr.* 53(5), 1900-1908.

Zhang, Y., Dijk, M.A.v., Liu, M., Zhu, G. and Qin, B. 2009 The contribution of phytoplankton degradation to chromophoric dissolved organic matter (CDOM) in eutrophic shallow lakes: Field and experimental evidence. *Water Res.* 43(18), 4685-4697

.

Chapter 6

INVESTIGATING THE IMPACT OF WATER TREATMENT ON
THE FLUORESCENCE SPECTRA OF HUMIC SUBSTANCES
IN SURFACE AND GROUND WATERS

Parts of this chapter are based on:

Baghoth, S.A., Maeng, S.K., Salinas Rodríguez, S.G., Ronteltap, M., Sharma, S., Kennedy, M. and Amy, G.L. 2008 An urban water cycle perspective of natural organic matter (NOM): NOM in drinking water, wastewater effluent, storm water, and seawater. *Water Sci. Technol. Water Supply* 8(6), 701-707.

Baghoth, S.A., Mosebolatan, K.O., Sharma, S.K. and Amy, G.L. Investigating the impact of water treatment on the fluorescence spectra of humic substances in surface and ground waters. Submitted to *Water Sci. Technol.* journal.

Summary

This study investigated the effects of different water treatment processes for the removal of natural organic matter (NOM) in surface and ground waters on the fluorescence characteristics of the NOM. The concept of online monitoring of dissolved organic carbon (DOC) concentration using fluorescence excitation-emission matrices (F-EEM) would generally be based on a fixed pair of excitation and emission wavelengths, such as for a humic-like peak or a protein-like peak. However, some treatment processes are known to result in a shift in the location of the fluorescence peaks and using the same pair of excitation emission wavelengths could potentially result in errors in the measurement of the maximum fluorescence intensity. This study focused on the spectral shifts of a humic-like peak (peak C), at an excitation wavelength in the visible region of 300-370 nm and an emission wavelength between 400 and 500 nm, and investigated the amount of error in the fluorescence intensity maximum if the shift in the location of peak C is not taken into account. Raw and treated surface and ground water samples were analyzed for F-EEM and the shift in the fluorescence spectra as well as the percentage error of the fluorescence intensity maximum of peak C were determined. The samples were treated for NOM removal in coagulation jar tests, pilot plants and full-scale water treatment plants. Coagulation of surface and ground water with iron chloride and alum resulted in a shift in the emission wavelength of humic-like peak C of between 8 and 18 nm, and an error in the maximum fluorescence intensity ranging between 2% and 6% if the shift is not taken into account. There was no significant difference in the spectral shift of peak C or in the error in the maximum fluorescence intensity between coagulation alone and coagulation followed by ozonation of ground water. NOM removal with ion exchange (IEX) alone generally resulted in a higher shift in peak C and a higher percentage error in the maximum fluorescence intensity than with coagulation, biological activated carbon (BAC) filtration or a combination of treatments. The impact of IEX treatment on the error of maximum fluorescence intensity was higher for surface than for ground waters, likely due to differences in molecular weight distribution of surface and ground water NOM. The results demonstrate that for NOM removal treatments other than IEX, the errors in the maximum fluorescence intensity that would result from ignoring the fluorescence spectral shifts are generally low (≤ 5%), and a fixed excitation emission wavelength pair for peak C could be used for online monitoring of NOM in water treatment plants. If IEX is included in the water treatment train, the resultant spectral shifts and fluorescence intensity errors should be ignored only if the process is located at the start of the treatment train.

6.1   Introduction

Natural organic matter (NOM) is a heterogeneous mixture of naturally occurring compounds found abundantly in natural waters. NOM originates from living and dead plants, animals and microorganisms, and from the degradation products of these sources (Chow et al., 1999). The concentration, composition and chemistry of NOM are highly variable and depend on the sources organic matter, the physicochemical properties of the water such as temperature, ionic strength and pH and the main cation components; the surface chemistry of sediment sorbents that act as solubility control; and the presence of photolytic and microbiological degradation processes (Leenheer and Croue, 2003). NOM in general significantly influences water treatment processes such as coagulation, oxidation, adsorption, and membrane filtration and some of its constituents are particularly problematic. In addition to aesthetic problems such as colour, taste and odour, it contributes to the fouling of membranes, serves as precursor for the formation of disinfection by-products, increases the exhaustion and usage rate of activation carbon and also certain fractions of NOM promotes microbial growth and corrosion in the distribution system (Amy, 1994; Owen et al., 1993).

The extent to which NOM affects water treatment processes depends on its quantity and physicochemical characteristics. NOM that is rich in aromatic structures such as carboxylic and phenolic functional groups have been found to be highly reactive with chlorine, thus forming forming DBPs (Reckhow et al., 1990). These aromatic structures are commonly present as a significant percentage of humic substances, which typically form over 50% of NOM. Hydrophobic and large molecular humic substances are enriched with aromatic structures and are readily removed by conventional drinking water treatment consisting of flocculation, sedimentation and filtration. In contrast, less aromatic hydrophilic NOM is more difficult to remove and is a major contributor of easily biodegradable organic carbon, which promotes microbiological regrowth in the distribution system. An understanding of the behaviour of different fractions or constituents of NOM present in water is crucial to understanding their fate and impact during water treatment and in water distribution systems. However, the heterogeneous nature of NOM makes it difficult to characterize in terms of structure and functional groups present.

Fluorescence spectroscopy is increasingly being used to characterize NOM, fractions of which   fluoresce when excited by UV and blue light. The fluorescence intensity and characteristics depend on, the concentration and composition of NOM, as well on other factors such as pH, temperature and ionic strength of the water. Fluorescence spectroscopy permits rapid data acquisition of aqueous samples at low natural concentrations. The relatively low expense and high sensitivity of fluorescence measurements, coupled with rapid data acquisition of water samples at low natural concentrations, have made fluorescence spectroscopy using fluorescence excitation-emission matrices (F-EEM) attractive for NOM characterization of water samples. F-EEM analysis is a fluorescence spectroscopy technique that is increasingly being used to characterize aquatic NOM (Chen et al., 2003;Wu et al., 2003; Coble et al., 1990; Coble et al., 1993; Mopper and Schultz, 1993). This characterization has typically involved the use of excitation-emission wavelength pairs to identify fluorophores based on the location of fluorescence peaks on F-EEM contour plots (Coble, 1996). These peaks have been used to distinguish between humic-like NOM, with longer emission wavelengths (> 350 nm), and protein-like NOM, with shorter emission wavelengths (≤ 350 nm). Other methods for NOM characterization using F-EEMs include: fluorescence regional integration (FRI) (Chen et al., 2003); multivariate data analysis (e.g. Principal Component Analysis, PCA, and Partial Least Squares regression, PLS) (Persson and Wedborg, 2001); and multi-way data analysis using parallel factor analysis (PARAFAC) (Stedmon et al., 2003). PARAFAC has been used to decompose F-EEMs into individual components some of which have been attributed to protein-like or humic-like NOM (Hunt and Ohno, 2007, Stedmon et al., 2003, Stedmon and Markager, 2005, Stedmon et al., 2007a, Stedmon et al., 2007b, Yamashita et al., 2008).

The traditional peak picking method involves the use of excitation-emission wavelength pairs to identify fluorophores based on the location of the maximum fluorescence intensity (Coble, 1996). The fluorescence intensity peaks are picked from a contour plot of F-EEMs and the excitation and emission wavelength pairs at which they occur are used to characterize the NOM fluorescence. A review of recent literature demonstrated the potential of F-EEMs as a successful monitoring tool for recycled water systems (Henderson et al., 2009) and it has been used to predict TOC removal during surface water treatment by clarification (Bieroza et al., 2009).

NOM fluorescence is commonly attributed to humic-like and protein-like fluorophores which have fluorescent signals with distinct locations of excitation and emission maxima (Mopper and Schultz, 1993; Coble, 1996). Protein-like fluorescence peaks occur at excitation and emission wavelengths similar to those of tryptophan (peak T) and tyrosine (peak B) amino acids. Humic-like fluorescence peaks occur at higher emission wavelengths. The general fluorescence properties of humic substances, which comprise most of the organic matter in most natural waters, are categorised by two maxima in their F-EEMs, one from excitation in the UV region at 250-260 nm and with emission maximum between 400 and 500 nm (peak A), and one from excitation in the visible region of 300-370 nm and with emission maximum between 400 and 500 nm (peak C) (Coble, 1996). Peak C fluorescence intensity has been demonstrated to correlate with TOC concentration (Bieroza et al., 2009), and its emission wavelength correlates with the molecular weight, aromaticity and the degree of hydrophobicity of the NOM (Coble, 1996; Wu et al., 2003; Baker et al., 2008).

This study investigates the effects of different water treatment processes for the removal of NOM in surface and ground waters on the fluorescence characteristics of peak C, which is representative of the humic fraction, the most abundant and reactive component of NOM. Potential online monitoring of DOC using F-EEM would generally be based on a fixed pair of excitation and emission wavelengths, such as peak C or peak T. However, some treatment processes are known to result in a shift in the location of the peaks and using the same pair of excitation emission wavelengths could potentially result in errors in the measurement of the maximum fluorescence intensity, particularly if online measurement is envisaged. This study investigates the amount of error in the fluorescence intensity maximum if the shift in the location of peak C is not taken into account.

6.2   Materials and methods

6.2.1 Sampling

For bench-scale coagulation tests, raw surface and ground water samples were used. The surface water samples were collected from River Meuse, Rotterdam, The Netherlands. Ground water samples were collected from water wells of Oldeholtpade drinking water treatment plant of Vitens Water Supply Company of North Holland, The Netherlands. Vitens is one of the water companies serving the provinces of Friesland, Overijssel, Gelderland and Flevoland. Oldeholtpade water treatment plant has only one well field of Oldeholtpade. The treatment process consists of plate aeration, rapid sand filtration, pellet softening, rapid sand filtration and ion exchange (IEX) using strong-base macro-porous acrylic anion resins (Purolite A860S). Samples were also collected from a pilot plant at Weesperkarspel drinking water treatment plant of Waternet, The Netherlands, which was set up to treat surface water for NOM removal using IEX resins. Surface and ground water samples were also collected from full-scale treatment plants, two of which are in France and four in The Netherlands.

6.2.2 Tests

6.2.2.1 Jar tests

The coagulation and flocculation experiments were carried out using a six-paddle jar tester. Six aliquots of the sample (1 L) were collected and, while stirring at 120 rpm for one minute to promote coagulation, the coagulant (iron chloride or aluminium sulphate (alum)) was dosed to five of the jars leaving the first one as a blank and the required amount of NaOH (0.1 M) or HCl (0.1 M) added to adjust the pH to the required level which was then recorded. The jars were then stirred at 45 rpm for 10 minutes to promote flocculation, and then allowed to settle for up to 1 hour before sampling. Samples of each of the jars were taken and filtered through a 0.45 μm pore size Whatman RC 55 regenerated cellulose membrane filters to remove particulates and each sample was analyzed for DOC, UVA_{254}, SUVA and fluorescence using F-EEMs. Details of the experimental setup, water quality data and results are given elsewhere (Mosebolatan, 2010).

6.2.2.2 Bench-scale Ozonation tests

Bench-scale ozonation tests were conducted using a batch reactor. Ozone was generated in the laboratory using a semi-batch an ozone generator (Trailigaz Ozonizer, LABO LO type) which employs the corona discharge method with dehumidified atmospheric air as the source of oxygen. The amount of ozone applied was determined from the generator mainly based on off-gas required volume and required time. Ozone was bubbled through 8 L of the water sample in a glass reactor vessel using a glass diffuser while continuously stirring in order to allow effective gas-liquid contact and mass transfer of ozone to the aqueous phase. Prior to ozonation of the samples, ozonation control tests were carried out to determine the transfer efficiency of ozone to the water sample and the efficiencies varied between 49% and 75%. The ozone doses applied ranged from 0.5 to 1.0 mg/mg O_3/DOC. After each ozonation experiment, a sample of the ozonated water was analyzed for dissolved organic carbon (DOC), ultraviolet absorbance at 254 nm (UVA_{254}), specific UVA_{254} (SUVA), which is calculated as a ratio of UVA_{254} and DOC, and F-EEMs. Details of the experimental setup, water quality data and results are given elsewhere (Mosebolatan, 2010).

6.2.3 Analytical methods

The raw water was analyzed for various water quality parameters, including pH, turbidity, electrical conductivity (EC), dissolved oxygen (DO), alkalinity and the presence and characteristics of NOM such as DOC, UVA_{254}, SUVA and F-EEM.

6.2.3.1 DOC and UVA_{254} measurements

All of the samples were pre-filtered through a 0.45 μm pore size Whatman RC 55 regenerated cellulose membrane filters, which were previously soaked overnight in Milli-Q® water in order to minimize leaching of DOC from the filter. DOC concentrations of all pre-filtered samples were determined by the catalytic combustion method using a Shimadzu TOC-V_{CPN} organic carbon analyzer. UVA_{254} of each sample was measured at room temperature (20±1°C) and ambient pH using a Shimadzu UV-2501PC UV-VIS scanning spectrophotometer. Measurements were made in duplicate and the average taken as the DOC concentration. Samples for UVA_{254} were adjusted to pH of 7.0 prior to measurement with a

Perkin Elmer (Lambda 20 1.11) spectrophotometer. The SUVA values were determined by dividing the UVA_{254} by the corresponding DOC concentration.

6.2.3.2   Fluorescence Excitation Emission Matrices (F-EEM)

Fluorescence intensities for all samples were measured at ambient pH and room temperature (20±1°C) using a FluoroMax-3 spectrofluorometer (Horiba Jobin Yvon). The pre-filtered samples were diluted to a DOC concentration of 1 mg C/L using ultrapure water obtained from a Milli-Q® water purification system prior to fluorescence measurements. F-EEMs were generated for each sample by scanning over excitation wavelengths between 240 and 450 nm at intervals of 10 nm and emission wavelengths between 290 and 500 nm at intervals of 2 nm. An F-EEM of Milli-Q® water was obtained and this was subtracted from that of each sample in order to remove most of the water Raman scatter peaks. Since samples were diluted to a DOC concentration of 1 mg C/L prior to measurements, each blank subtracted F-EEM was multiplied by the respective dilution factor and Raman-normalized by dividing by the integrated area under the Raman scatter peak (excitation wavelength of 350 nm) of the corresponding Milli-Q® water, and the fluorescence intensities reported in Raman units (RU).

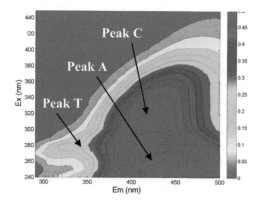

Figure 6.1 F-EEM of raw surface water showing fulvic-like fluorescence (peak C), humic-like fluorescence (Peak A) and tryptophan-like fluorescence (Peak T).

Figure 6.1 is a typical source water F-EEM contour plot for the Weesperkarspel water treatment plant showing the locations of the fluorescence intensity of protein-like peak B and humic-like peaks A and C. The maximum intensities of the two humic-like peaks A and C are much higher than of the protein-like peak T, indicating that the source water NOM is predominantly humic in character. The F-EEMs of all the water samples analyzed in this study were characterized by the presence of these three fluorescence peaks.

6.3   Results and discussion

6.3.1 Jar tests with surface water and ground water

6.3.1.1   Raw water characteristics

Surface water and ground water samples were used in coagulation jar tests and in coagulation followed by ozonation tests. Surface water samples were collected from River Meuse between November 6, 2009, and February 16, 2010, and ground water samples were collected from Oldeholtpade water treatment plant of Vitens Water Supply Company on the 6th of January 2010. Mean values of water quality parameters of raw water samples from River Meuse and Oldeholtpade ground water treatment plant are shown in Table 6.1.

Table 6.1 Average water quality parameters of raw water samples from River Meuse and Oldeholtpade ground water treatment plant.

Water quality parameter	Units	River Meuse water	Oldeholtpade source water
Temperature	^{0}C	20	20
Turbidity	NTU	16	112
pH		7.9	6.80
Electrical conductivity (EC)	µS/cm	1265	418
Alkalinity	mg/L $CaCO_3$	101	375
DO	mg/L	9.9	
Total iron	mg/L		9.8
DOC	mg C/L	3.9	8.1
UVA$_{254}$	1/cm	0.098	0.318
SUVA	L/(mg-m)	2.5	3.9

The mean DOC concentration of water samples from the River Meuse (3.9 mg C/L) is within the typical range for river waters. The corresponding mean SUVA value of 2.5 L/(m-mg) is indicative of NOM enriched in non-humic substances which are relatively hydrophilic, less aromatic and of lower molecular weight compared to waters with higher SUVA values (Edzwald, 1993; Edzwald and Tobiason, 1999). The mean DOC concentration of Oldeholtpade ground water (3.9 mg C/L) is much higher than is typical of ground waters (< 2 mg C/L). The mean SUVA value of about 4 L/(m-mg) is indicative of NOM which is a mixture of aquatic humics and other NOM, a mixture of hydrophobic and hydrophilic NOM and a mixture of molecular weights (Edzwald et al., 1985).

6.3.1.2 Effect of coagulation pH and coagulant dose on DOC removal by coagulation

Coagulation jar tests with iron chloride and alum coagulants at different pH were conducted in order to select an optimum coagulation pH and coagulant dose for DOC removal. These optimum conditions were then used in coagulation jar tests to investigate the effect of coagulation on NOM fluorescence of surface water and ground water. The coagulation experiment were conducted using surface water samples from River Meuse and ground water samples from Oldeholtpade well field at a coagulation pH of 5.5, 6.0, 6.5 and 7.0, and iron chloride and alum coagulants doses of 5, 10, 15, 20 and 30 mg/L (as Fe or Al). Figures 6.2 illustrates the residual DOC for different doses of iron chloride and alum at different coagulation pH for both River Meuse and Oldholtpade water samples.

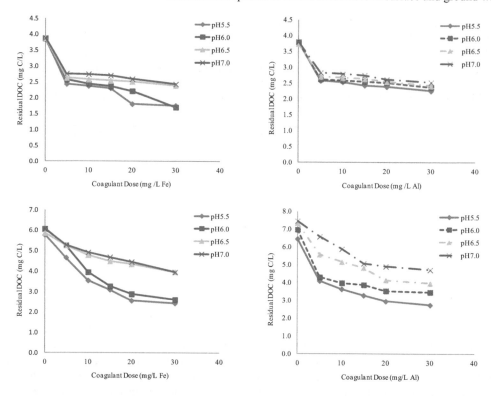

Figure 6.2 Coagulation of River Meuse water (top) and Oldeholtpade ground water (bottom) with FeCl₃ (left) and alum (right).

For both River Meuse water and Oldeholtpade ground water, the removal efficiency of DOC increased with increasing coagulant dose, irrespective of the type of coagulant. For both iron chloride and alum, the removal efficiency was higher at lower pH values and the highest removal was achieved at a pH of 5.5 for both River Meuse water and Oldeholtpade ground water. This is consistent with other studies showing that the optimum coagulation pH values are 5.5 for iron chloride (Vilg-Ritter et al., 1999; Freese et al., 2001). At this pH, DOC removal starts to level off at a coagulant dose of 20 mg/L for both iron chloride and alum. At a coagulant dose of 20 mg/L, the DOC removal for River Meuse water and Oldehotpade ground water samples were 53% and 56%, respectively, using iron chloride, and 37% and 54%, respectively, using alum. Iron chloride demonstrated a better DOC removal efficiency than alum for both waters. Other studies have also found that iron coagulants generally performed better for removal of NOM than alum (Vilg-Ritter et al., 1999; Volk et al., 2000).

6.3.1.3 Fluorescence spectra of surface and ground waters in coagulation jar tests

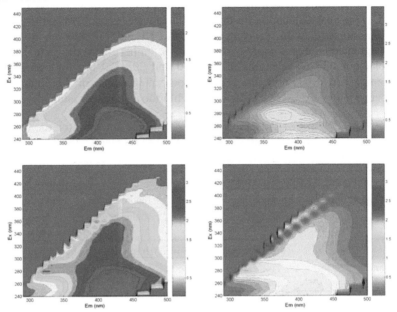

Figure 6.3 F-EEM spectra of River Meuse water before (left) and after (right) coagulation with iron chloride (top) and alum (bottom) at a dose of 20 mg/L and coagulation pH of 5.5.

Coagulation jar test experiments were carried out on River Meuse surface and Oldeholtpade ground waters using increasing doses of iron and alum coagulants at the optimised coagulation pH of 5.5. Samples of the coagulated waters were then analysed for F-EEM in order to understand the impact of coagulation treatment on fluorescence peak locations. In conventional drinking water treatment involving biological activated carbon (BAC) filtration, ozonation is typically carried out after coagulation/flocculation and prior to BAC filtration. Ozonation has been shown to breakdown NOM into smaller molecular weight compounds, which results in the shift of the fluorescence spectra to shorter emission wavelengths. The combined effect of treatment by coagulation followed by ozonation on the fluorescence peak locations was therefore investigated. The investigation of the changes in the fluorescence peak locations was restricted to the humic-like peak C, which has been shown to be abundant in aquatic NOM. The F-EEM spectra of River Meuse surface water before and after coagulation with iron chloride and alum at a coagulation pH of 5.5 and a coagulant dose of 20 mg/L are presented in Figure 6.3.

Because of the broad fluorescence peak C for the River Meuse water, the effect of coagulation on its location is not visually apparent from the F-EEM contour plots. To investigate the shifting of peak C during coagulation, the traditional peak-picking method was employed, which involved manually selecting the point of maximum intensity fluorescence intensity within the peak C region. This procedure was performed using Matlab, which simplified the process and permitted quick data acquisition.   The results of the analysis of the impact of coagulation with iron chloride and alum, at a dose of 20 mg/L and a coagulation pH of 5.5, on peak C for River Meuse NOM are presented in Table 6.2. The emission and excitation wavelengths ranged between 410 and 414 nm and 310 and 320 nm, respectively, in the raw water, and between 398 and 400 nm and 310 nm, respectively, in the coagulated water. For both iron chloride and alum, there was a slight shift towards shorter emission wavelengths, indicating that coagulation preferentially removed larger components of NOM. If the peak C location for the coagulated water is taken to be the same as for the raw water (emission/excitation wavelengths of 414/310 and 410/320 nm for iron chloride and alum coagulated water samples, respectively), rather than the actual peak C location (emission/excitation wavelengths of 400/310 and 398/310 nm for iron chloride and alum water samples, respectively), the resulting percentage error in the value of the maximum fluorescence intensity would be 6% for iron chloride and 3% for alum.

Table 6.2 The impact of coagulation with iron chloride and alum on the fluorescence spectra of River Meuse NOM. "Arbitrary Peak C" is the fluorescence intensity at excitation/emission wavelength of the maximum fluorescence intensity for the raw water.

Sample	Actual Peak C			Arbitrary Peak C	
	Emission wavelength (nm)	Excitation wavelength (nm)	Fluorescence intensity (RU)	Fluorescence intensity (RU)	Percentage error (%)
Coagulation of River Meuse water with iron chloride					
Raw water	414	310	1.928		
Coagulated water	400	310	0.356	0.335	6
Coagulation of River Meuse water with alum					
Raw water	410	320	2.538		
Coagulated water	398	310	0.571	0.556	3

The F-EEM spectra of Oldeholtpade ground water before and after coagulation with iron chloride and alum at a coagulation pH of 5.5 and a coagulant dose of 30 mg/L are presented in Figure 6.4. The dominance of the humic-like fluorescence is more apparent than for River Meuse surface water. The results of the analysis of the impact of coagulation with iron chloride and alum on peak C for Oldeholtpade ground water are presented in Table 6.3. The emission and excitation wavelengths were 436 and 320 nm, respectively, in the raw water. The emission wavelength ranged between 418 and 428 nm and the excitation wavelength was 320 nm in the coagulated water. As in the case of River Meuse surface water, there was, for both iron chloride and alum, a slight shift towards shorter emission wavelengths. If the peak C location for the coagulated water is taken to be the same as for the raw water (emission/excitation wavelengths of 436/320 nm), rather than the actual peak C location (emission/excitation wavelengths of 428/320 and 418/320 nm for iron chloride and alum, respectively), the resulting percentage error in the value of the maximum fluorescence intensity would be 2% for iron chloride and 3% for alum.

In order to analyse the effect of coagulation followed by ozonation on fluorescence characteristics of surface and ground water NOM, coagulation jar tests were conducted. Coagulation of Oldeholtpade ground water was carried out at a pH of 5.5 and a coagulant dose of 20 mg/L for both iron chloride and alum. Ozonation of the supernatant water was then carried out using O_3/DOC ratio of 1 mg/L: 1 mg/L. The combined effect of coagulation with metal coagulants followed by ozonation on the fluorescence peak locations was then investigating by analyzing the shift in the location of the peak C. F-EEMs were generated for the water samples before coagulation and after the combined treatment of coagulation and ozonation. F-EEM spectra of Oldeholtpade ground water before coagulation, after coagulation with iron chloride followed by ozonation and after coagulation with alum followed by ozonation are presented in Figure 6.5.

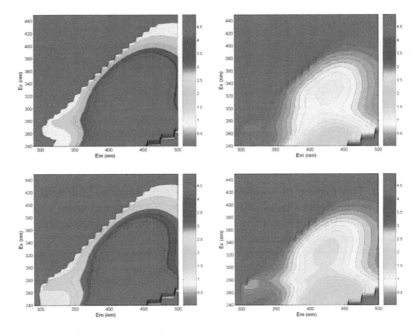

Figure 6.4 F-EEM spectra of Oldeholtpade ground water before (left) and after (right) coagulation with iron chloride (top) and alum (bottom) at a dose of 30 mg/L and pH of 5.5.

Table 6.3 The impact of coagulation with iron chloride and alum on the fluorescence spectra of Oldeholtpade ground water NOM. "Arbitrary Peak C" is the fluorescence intensity at excitation/emission wavelength of the maximum fluorescence intensity for the raw water.

Sample	Actual Peak C			Arbitrary Peak C	
	Emission wavelength (nm)	Excitation wavelength (nm)	Fluorescence intensity (RU)	Fluorescence intensity (RU)	Percentage error (%)
Coagulation of Oldeholtpade ground water with iron chloride					
Raw water	436	320	9.514		
Coagulated water	428	320	1.128	1.110	2
Coagulation of Oldeholtpade ground water with alum					
Raw water	436	320	7.953		
Coagulated water	418	320	1.362	1.319	3

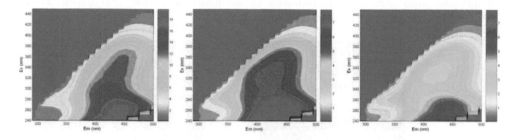

Figure 6.5 F-EEM spectra of Oldeholtpade ground water before coagulation (left), after coagulation with iron chloride followed by ozonation (middle) and after coagulation with alum followed by ozonation (right).

Table 6.4 The impact of coagulation with iron chloride, and with alum, followed by ozonation on the fluorescence spectra of Oldeholtpade ground water NOM. "Arbitrary Peak C" is the fluorescence intensity at excitation/emission wavelength of the maximum fluorescence intensity for the raw water.

Sample	Actual Peak C			Arbitrary Peak C	
	Emission wavelength (nm)	Excitation wavelength (nm)	Fluorescence intensity (RU)	Fluorescence intensity (RU)	Percentage error (%)
Coagulation/ozonation of Oldeholtpade ground water water with iron chloride and alum					
Raw water	436	330	14.014		
Coagulated/ozonated water (iron chloride)	426	330	7.510	7.270	3
Coagulated/ozonated water (alum)	418	340	4.32	4.203	3

The results of the analysis of the impact of combined treatment by coagulation and ozonation on the location of peak C for Oldeholtpade ground water are presented in Table 6.4. As in the case of coagulation alone, there was, for both iron chloride and alum, a slight shift towards shorter emission wavelengths. If the peak C location for the coagulated water is taken to be the same as for the raw water (emission/excitation wavelengths of 436/320 nm), and represented as arbitrary peak C in the table, rather than the actual peak C location (emission/excitation wavelengths of 428/320 and 418/320 nm for iron chloride and alum, respectively), the resulting percentage error in the value of the maximum fluorescence intensity would be 3% for both iron chloride and alum. There is almost no difference in the percentage error in the value of the fluorescence intensity of peak C between treatment by coagulation alone and by combined treatment of coagulation and ozonation. The peak shift towards shorter emission wavelengths (blue shifting) is a result of the intact removal, by coagulation, of larger molecular weight NOM, which emit at longer wavelengths. Ozonation also results in blue shifting of fluorescence peaks, a reflection of the transformation of larger to lower molecular weight NOM, as well as quenching of the fluorophores, a reflection of the transformation from more to less aromatic NOM and/or the changes in the functional groups.

6.3.2 Fluorescence spectral across ion exchange pilot plant

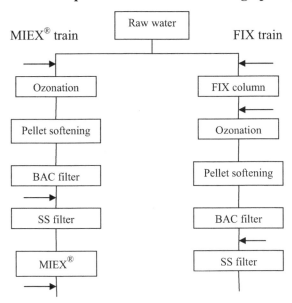

Figure 6.6 Process schemes of the fluidized bed ion exchange (FIX) and Magnetic ion exchange resins (MIEX®) pilot plant trains. The arrows show the sampling points.

Figure 6.6 shows the process schemes of the pilot plant at Weesperkarspel drinking water treatment plant of Waternet, The Netherlands, which was set up to treat surface water for NOM removal using IEX resins. The first treatment train involved treatment by magnetic ion exchange resins (MIEX®), while the second involved treatment by fluidized bed ion-exchange (FIX). Figure 6.7 illustrates the F-EEM spectra of Weesperkarspel source water before and after treatment by FIX and BAC filtration, in the FIX pilot plant train, and by BAC filtration and MIEX®, in the MIEX® pilot plant train. Table 6.5 presents the results of the analysis of the impact of IEX pilot plant treatment on the location of peak C for surface water of Weesperkarspel treatment plant. Similar to coagulation, FIX and MIEX® treatments resulted in a slight shift towards shorter emission wavelengths. A comparison of the F-EEM of BAC filtered samples of the street with and without FIX showed that the shift to shorter emission wavelengths after FIX is not propagated through the subsequent treatment by ozonation, pellet softening and BAC filtration. Whereas FIX reduced the emission of wavelength of the raw water from 417 to 399 nm, the emission wavelengths of BAC filtered water was nearly the same for the street with (419 nm) and without (418 nm) FIX. If the peak C location for the raw water is used as the default for all the samples, the resulting percentage error in the value of the maximum fluorescence intensity is significantly higher with FIX treatment alone than when combined with subsequent treatment by ozonation, pellet softening and BAC filtration. Because MIEX® is at the end of the treatment train, a similar observation cannot be made. However, since both FIX and MIEX® use anion exchange resins, it is likely that a similar result would be obtained for the latter if it were upstream of the other treatment processes. The percentage error in the value of the maximum fluorescence intensity for MIEX® (13%) is more than twice that for FIX, which could be a result of the difference in their locations along their respective treatment trains, with the latter at the beginning and the former at the end of the treatment train.

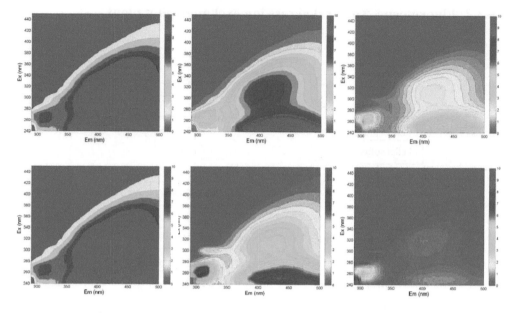

Figure 6.7 F-EEM spectra of Weesperkarspel source water before (left) and after treatment by fluidized bed ion-exchange (FIX) (top middle) and BAC filtration (top right), in the FIX pilot plant street, and by BAC filtration (bottom middle) and MIEX® (bottom right), in the MIEX® pilot plant street.

Table 6.5 The impact of pilot plant treatment with ion exchange resins using fluidized bed ion-exchange (FIX) and magnetic ion-exchange resins (MIEX®) on Weesperkarspel source water NOM fluorescence spectra. "Arbitrary Peak C" is the fluorescence intensity at excitation/emission wavelength of the maximum fluorescence intensity for the raw water.

	Ion Exchange Pilot Plant				
	Actual Peak C			Arbitrary Peak C	
Sample	Emission wavelength (nm)	Excitation wavelength (nm)	Fluorescence intensity (RU)	Fluorescence intensity (RU)	Percentage error (%)
Raw water	417	307	18.7		
FIX effluent	399	310	7.7	7.2	6 ± 1.0
BAC effluent (FIX)	419	313	1.9	1.9	2 ± 0.9
BAC effluent (MIEX®)	418	313	4.0	3.9	3 ± 1.8
MIEX® effluent	401	323	0.8	0.7	13 ± 6.4

6.3.3 Fluorescence spectra across full-scale surface and ground water treatment plants

The impact of different water treatment processes of six full-scale surface and ground water treatment plants on NOM fluorescence spectra was also investigated. Two of these treatment plants, Choisy-le-Roi and Neuilly-sur-Marne, are surface water treatment plants of Syndicat des Eaux d'Ile de France (SEDIF), which supply drinking water to the suburbs of the city of Paris, France. Descriptions of the two plants, the sampling details and other water quality parameters are given in Chapter 5. One of the treatment plants, Oldeholtpade, treats ground water and a description of the treatment processes, details of the water quality data and NOM characterization results are given elsewhere (Mwesigwa, 2007). Samples from two water treatment plants of Waternet, a pre-treatment plant at Loenderveen and a post treatment plant at Weesperkarspel, were also analysed for F-EEM. Details of these two treatment plants and the water quality data of the samples collected are presented in Chapter 3. The sixth treatment plant, Andijk water treatment plant of PWN Water Supply Company of North-Holland, The Netherlands, treats surface water from Lake Ijssel. At the time of the study, Andijk had installed a pilot plant comprising MIEX® and ultrafiltration and this study focused only on the MIEX® treatment. Details of the water quality data and NOM characterization results for the Andijk pilot plant are given elsewhere (Tesoura, 2008).

Table 6.6 The impact of different water treatment processes of several full-scale plants on NOM fluorescence spectra. "Arbitrary Peak C" is the fluorescence intensity at excitation/emission wavelength of the maximum fluorescence intensity for the raw water.

Sample	Actual Peak C			Arbitrary Peak C	
	Emission wavelength (nm)	Excitation wavelength (nm)	Fluorescence intensity (RU)	Fluorescence intensity (RU)	Percentage error (%)
Andijk water treatment plant					
Raw water	417	303	7.92		
MIEX® effluent	393	307	1.83	1.64	9 ± 9.4
Oldehaltpade water treatment plant					
Raw water	450	330	1.45		
IEX effluent	436	330	0.99	0.94	5
Choisy-le-Roi water treatment plant					
Raw water	421	310	0.527		
Settled water	420	310	0.291	0.287	2 ± 1.8
Treated water	419	307	0.084	0.082	2 ± 2.2
Neuilly-sur-Marne water treatment plant					
Raw water	429	327	0.585		
Settled water	423	327	0.392	0.388	1 ± 1.1
Treated water	420	320	0.056	0.054	4 ± 0.9
Loenderveen water treatment plant					
Raw water	437	320	1.520		
Coagulated water	435	322	1.160	1.154	1 ± 0.7
Weesperkarspel water treatment Plant					
Pretreated water	422	312	0.823		
Treated water	426	318	0.088	0.084	4 ± 3.0

The impact of different water treatment processes of the five full-scale plants and the MIEX® pilot plant on NOM fluorescence spectra was investigated using the location of peak C and the results are shown in Table 6.6. The location of peak C for the raw water was used as the default for all the samples and the resulting percentage error in the value of the maximum fluorescence intensity was determined by comparison with the actual location of peak C found using the peak-picking method. Similar to the results of the IEX pilot plant of Weesperkarspel treatment plant, treatment of surface water in the MIEX® pilot plant of Andijk resulted in a slight shift to a shorter emission wavelength. The resulting percentage error (9%) in the value of the maximum fluorescence intensity is comparable to that of Weesperkarspel pilot plant (13%). These results show that IEX results in a higher percentage error than does coagulation or combined treatment.

The overall treatment did not significantly change the fluorescence spectra for all of the full-scale treatment plants, except the one of Oldeholtpade, which has IEX as the last treatment process. For these treatment plants, the calculated percentage errors were generally less than 5%. For Oldeholtpade treatment plant, the shift in peak C location and the percentage errors of the maximum fluorescence intensity after IEX treatment are is similar to that of the results of the FIX treatments. The results imply a higher effect on the fluorescence spectra by MIEX® than by the other IEX treatments. The results also indicate that for all the full-scale treatment plants, if the shift in peak C is ignored, the overall treatment generally results in errors of not more than 5% in the maximum fluorescence intensity.

6.4   Conclusions

Based on the results of the analysis of F-EEM of surface and ground waters during coagulation jar tests, pilot plant water treatment and full-scale water treatment, the following conclusions about the impact of different water treatment processes on the spectral shift of fluorescence humic-like peak C can be made:

•   Coagulation of surface water with iron chloride and alum resulted in a shift in the emission wavelength of humic-like peak C of between 12 and 15 nm, and an error in the maximum fluorescence intensity of 6% for iron chloride and 3% for alum if the shift is not taken into account.

•   Coagulation of ground water with iron chloride and alum resulted in a shift in the emission wavelength of humic-like peak C of between 8 nm (iron chloride) and 18 nm (alum), and an error in the maximum fluorescence intensity of 2% for iron chloride and 3% for alum if the shift is not taken into account. The significant differences in the spectral shift between iron chloride and alum could be due to the effect of the high iron content of the raw ground water.

•   There was no significant difference in the spectral shift of peak C or in the error in the maximum fluorescence intensity between coagulation alone or coagulation followed by ozonation of ground water.

•   NOM removal with IEX (using FIX, MIEX® or other IEX resins) alone generally resulted in a higher shift (up to -24 nm for Em and 4 nm for Ex) in peak C and a higher percentage error (up to 13%) in the maximum fluorescence intensity than with coagulation, BAC filtration or a combination of treatments.

•   The impact of IEX treatment on the error of maximum fluorescence intensity was higher for surface than for ground waters, likely due to differences in molecular weight distribution of surface and ground water NOM.

•   The results demonstrate that for NOM removal treatments other than IEX, the errors in the maximum fluorescence intensity that would result from ignoring the fluorescence spectral shifts are generally low (≤ 5%), and a fixed excitation emission wavelength pair for peak C could be used for online monitoring of NOM in water treatment plants. If IEX is included in the treatment train, then its location should be upstream in order to minimize the resultant spectral shifts and fluorescence intensity errors.

6.5   References

Amy, G. (1994) Using NOM Characterisation for Evaluation of Treatment.In *Proceedings of Workshop on "Natural Organic Matter in Drinking Water, Origin, Characterization and Removal"*, September 19–22, 1993, Chamonix, France. American Water Works Association Research Foundation, Denver, USA, p. 243.

Baker, A., Tipping, E., Thacker, S.A. and Gondar, D. 2008 Relating dissolved organic matter fluorescence and functional properties. *Chemosphere* 73(11), 1765-1772.

Bieroza, M., Baker, A. and Bridgeman, J. 2009 Relating freshwater organic matter fluorescence to organic carbon removal efficiency in drinking water treatment. *Sci. Total Environ.* 407(5), 1765-1774.

Chen, W., Westerhoff, P., Leenheer, J.A. and Booksh, K. 2003 Fluorescence Excitation-Emission Matrix Regional Integration to Quantify Spectra for Dissolved Organic Matter. *Environ. Sci. Technol.* 37, 5701-5710.

Chow, C.W.K., van Leeuwen, J.A., Drikas, M., Fabris, R., Spark, K.M. and Page, D.W. 1999 The impact of the character of natural organic matter in conventional treatment with alum. *Water Sci. Technol.* 40(9), 97-104.

Coble, P.G., Green, S.A., Blough, N.V. and Gagosian, R.B. 1990 Characterization of dissolved organic matter in the Black Sea by fluorescence spectroscopy. *Nature* 348, 432-435.

Coble, P.G., Schultz, C.A. and Mopper, K. 1993 Fluorescence contouring analysis of DOC Intercalibration Experiment samples: a comparison of techniques. *Marine Chemistry* 41, 173-178.

Coble, P.G. 1996 Characterization of marine and terrestrial DOM in seawater using excitation-emission matrix spectroscopy. *Marine Chemistry* 51, 325-346.

Edzwald, J.K., Becker, W.C. and Wattier, K.L. 1985 Surrogate parameter for monitoring organic matter and THM precursors. *J. Am. Water Works Assoc.* 77, 122-132.

Edzwald, J.K. 1993 Coagulation in drinking water treatment: particles, organics and coagulants *Water Sci. Technol.* 27(11), 21-35.

Edzwald, J.K. and Tobiason, J.E. 1999a Enhanced coagulation: US requirements and a broader view. *Water Sci. Technol.* 40(9), 63-70.

Edzwald, J.K. and Tobiason, J.E. 1999b Enhanced Coagulation:USA requirements and a broader View *Water Science and Technology* 40(9), 63-70.

Freese, S.D., Nozaic, D.J., Pryor, M.J., Rajogopaul, R., Trollip, D.L. and Smith, R.A. 2001 Enhanced coagulation: a viable option to advance treatment technologies in the South African context. *Water Science and Technology: Water Supply* 1(1), 33–41.

Henderson, R.K., Baker, A., Murphy, K.R., Hambly, A., Stuetz, R.M. and Khan, S.J. 2009 Fluorescence as a potential monitoring tool for recycled water systems: A review. *Water Res.* 43 863-881.

Hunt, J.F. and Ohno, T. 2007 Characterization of fresh and decomposed dissolved organic matter using excitation-emission matrix fluorescence spectroscopy and multiway analysis. *J. Agricultural and Food Chemistry* 55(6), 2121-2128.

Leenheer, J.A. and Croue, J.-P. 2003 Characterizing Dissolved Aquatic Organic matter: Understanding the unknown structures is key to better treatment of drinking water. *Environ. Sci. Technol.* 37(1), 19A-26A.

Mopper, K. and Schultz, C.A. 1993 Fluorescence as a possible tool for studying the nature and water column distribution of DOC components. *Marine Chemistry* 41, 229-238.

Mosebolatan, K.O. (2010) Changes in the physicochemical properties of natural organic matter (NOM) during drinking water treatment. MSc, UNESCO-IHE Institute for Water Education, Delft.

Mwesigwa, J.K. (2007) Relating NOM characteristics to treatability of groundwater: A case study of Vitens water supply company water treatment plants. MSc, UNESCO-IHE Institue for Water Education, Delft.

Owen, D.M., Amy, G.L. and Chowdhary, Z.K. (eds) (1993) Characterization of Natural Organic Matter and its Relationship to Treatability, American Water Works Association Research Foundation, Denver, CO.

Persson, T. and Wedborg, M. 2001 Multivariate evaluation of the fluorescence of aquatic organic matter. *Analytica Chimica Acta* 434(2), 179-192.

Reckhow, D.A., Singer, P.C. and Malcolm, R.L. 1990 Chlorination of Humic Materials - by-Product Formation and Chemical Interpretations. *Environ. Sci. Technol.* 24(11), 1655-1664.

Stedmon, C.A., Markager, S. and Bro, R. 2003 Tracing dissolved organic matter in aquatic environments using a new approach to fluorescence spectroscopy. *Marine Chemistry* 82, 239–254.

Stedmon, C.A. and Markager, S. 2005 Resolving the variability in dissolved organic matter fluorescence in a temperate estuary and its catchment using PARAFAC analysis. *Limnol. Oceanogr.* 50(2), 686-697.

Stedmon, C.A., Markager, S., Tranvik, L., Kronberg, L., Slätis, T. and Martinsen, W. 2007a Photochemical production of ammonium and transformation of dissolved organic matter in the Baltic Sea. *Marine Chemistry* 104, 227–240.

Stedmon, C.A., Thomas, D.N., Granskog, M., Kaartokallio, H., Papadimitriou, S. and Kuosa, H. 2007b Characteristics of dissolved organic matter in Baltic coastal sea ice: Allochthonous or autochthonous origins? *Environ. Sci. Technol.* 41(21), 7273-7279.

Tesoura, M.R. (2008) Relating Natural Organic Matter (NOM) Characteristics to Biostability. MSc, UNESCO-IHE Institue for Water Education, Delft.

Vilg-Ritter, A., Masion, A., Boulange, T., Rybacki, D. and Bottero, J.Y. 1999 Removal of natural organic matter by coagulation-flocculation: A pyrolysis-GC-MS study. *Environ. Sci. & Technol.* 33(17), 3027-3032.

Volk, C., Bell, K., Ibrahim, E., Verges, D., Amy, G. and Lechevallier, M. 2000 Impact of Enhanced and Optimized Coagulation on Removal of Organic Matter and its Biodegradable Fraction in Drinking Water. *Water Res.* 34(12), 3247-3257.

Wu, F.C., Evans, R.D. and Dillon, P.J. 2003 Separation and Characterization of NOM by High-Performance Liquid Chromatography and On-Line Three-Dimensional Excitation Emission Matrix Fluorescence Detection. *Environ. Sci. Technol.* 37, 3687-3693.

Yamashita, Y., Jaffe′, R., Maie, N. and Tanoue, E. 2008 Assessing the dynamics of dissolved organic matter (DOM) in coastal environments by excitation emission matrix fluorescence and parallel factor analysis (EEM-PARAFAC). *Limnol. Oceanogr.* 53(5), 1900-1908.

Chapter 7

MODELLING AND PREDICTION OF THE REMOVAL OF NOM AND FORMATION OF TRIHALOMETHANES IN DRINKING WATER TREATMENT

Parts of this chapter are based on:

Baghoth, S.A., Sharma, S.K. and Amy, G.L. Modelling and prediction of the removal of NOM and formation of trihalomethanes in drinking water treatment. Submitted to *Water Sci. Technol.* journal.

Summary

This study investigates the incorporation of fluorescence measurements, which have relatively low expense and high sensitivity and can be relatively cheaply installed for online measurements, to improve the monitoring of concentrations of dissolved organic carbon (DOC) and formation of total trihalomethanes (THMs) in a drinking water treatment. Florescence measurements were based on fluorescence excitation-emission matrices (F-EEMs), which involve the use of excitation-emission wavelength pairs to identify fluorophores (fluorescent NOM fractions) based on the location of fluorescence peaks on F-EEM contour plots (Coble, 1996). Predictive models are developed for the removal of NOM and formation of THMs after chlorine disinfection in a full-scale drinking water treatment plant (WTP) using several water quality parameters which were measured as part of a study to characterize NOM and it's removal during water treatment. Statistical methods using simple linear regression and stepwise multiple linear regression (MLR) were used to model and predict the concentrations of DOC and THMs in the treated water.

The source water DOC concentration could be moderately predicted ($r^2 = 0.58$) using a multiple linear regression relationship that included temperature, conductivity and turbidity. Whereas the use of PARAFAC fluorescence components slightly improved the prediction of finished water DOC concentration, the prediction accuracy was generally low for both simple linear and multiple linear regressions. The applied coagulation dose could be predicted ($r^2 = 0.91$, $p < 0.001$) using multiple linear regressions involving temperature, UVA_{254}, total alkalinity, turbidity and protein-like peak T fluorescence. The total THMs concentration of the finished water could be predicted ($r^2 = 0.88$, $p < 0.001$) using temperature, turbidity, ozone dose, UVA_{254} and fluorescence peaks T and M. However, when fluorescence peaks T and M were replaced with the five PARAFAC components, the resulting model, which involved temperature, turbidity, ozone dose, UVA_{254} and PARAFAC components C1 and C2, had a slightly reduced prediction accuracy ($r^2 = 0.74$, $p < 0.001$) for total THMs in the finished water.

Predictive modelling provides an alternative to relatively complex analytical methods for determining the concentrations of THMs in finished drinking water, which could be more expensive and time consuming. Whereas the models for predictions of THMs concentrations show good predictability for the specific treatment plant investigated, they cannot reliably be globally applied to other water utilities without further research. The reliability of the models could be further investigated through the collection of more data which would permit cross-validation of the models using data splitting methods or by using the newly collected as the validation data set and the old data as the model development (training) data set.

7.1   Introduction

Naturally occurring aquatic organic matter (NOM) has attracted significant interest in drinking water treatment because of its impact on water treatment processes such as coagulation, oxidation, adsorption, and membrane filtration. NOM affects drinking water quality in a number of ways: it contributes to formation of potentially carcinogenic disinfection by-products (DBPs) (Sharp et al., 2004), promotes biological regrowth in the water distribution system and contributes to colour, tastes and odours. The extent to which NOM affects water treatment processes depends on its quantity and physicochemical characteristics. NOM is a heterogeneous mixture of compounds found abundantly in natural waters and originates from living and dead plants, animals and microorganisms, and from the degradation products of these sources (Chow et al., 1999). NOM that is rich in aromatic structures such as carboxylic and phenolic functional groups have been found to be highly reactive with chlorine, thus forming DBPs (Reckhow et al., 1990a). These aromatic structures are commonly present as a significant percentage of humic substances, which typically form over 50% of NOM. Hydrophobic and large molecular weight humic substances are enriched with aromatic structures and are readily removed by conventional drinking water treatment consisting of flocculation, sedimentation and filtration. In contrast, less aromatic hydrophilic NOM is more difficult to remove and is a major contributor of easily biodegradable organic carbon, which promotes microbiological regrowth in the distribution system.

Chlorine is widely applied for disinfection of drinking water. It is a very effective disinfectant for inactivation of microbial organisms and provides residual protection against microbiological growth in drinking water distribution systems. Whereas there are several disinfectants in use today, chlorine remains the most inexpensive and widely used disinfectant in drinking water treatment (Chowdhury et al., 2007; Clark et al., 1998). However, a major disadvantage for the use of chlorine for disinfection is the formation of a variety of halogenated DBPs when it reacts with NOM in water during water treatment and/or in the water distribution system. The most prominent of these DBPs are the potentially carcinogenic trihalomethanes (THMs) and haloacetic acids (HAAs) (Richardson et al., 2002). In order to minimise consumers' exposure to hazardous DBPs while at same time maintaining the microbial quality of drinking water through adequate disinfection, THMs and HAAs are regulated in the United States by the Environmental Protection Agency, which allows maximum contaminant levels of 80 and 60 µg/L, respectively, as well as in several other countries such as Canada and the European Union.

The formation of THMs in drinking water depends on a number of factors such as the concentration of NOM as measured by DOC concentration, NOM character (particularly its aromaticity), chlorine dose, temperature, pH, concentration of bromide and ammonia, and reaction time (Richardson et al., 2002; Reckhow et al., 1990b; Peters et al., 1980). A significant amount of research has been carried out to develop models for formation of THMs in drinking water treatment using kinetics and statistical methods (Amy et al., 1987; Westerhoff et al., 2000). In an investigation of the formation and occurrence of THMs in a drinking water treatment plant involving chlorine disinfection, twenty-three water treatment parameters were measured and, using multivariate statistical methods, the following parameters were found to be among the most important for THM formation: water temperature, total organic carbon (TOC), chlorine dose, ultraviolet absorbance at 254 nm (UVA_{254}) and turbidity (Platikanova et al., 2007). In the same study, the concentration of total THMs, which represents the sum of the concentrations of the four THMs, chloroform ($CHCl_3$), bromodichloromethane ($CHCl_2Br$), chlorodibromo-methane ($CHBr_2Cl$) and bromoform ($CHBr_3$), was found to provide better prediction than that of the individual THMs (Platikanova et al., 2007). Surrogate parameters such as specific UVA_{254} (SUVA) and differential UV absorbance at 272 nm have also been used to predict the formation of THMs in drinking water (Ates et al., 2007; Korshin et al., 2002). NOM fluorescence is another spectroscopic property which has been used to investigate chlorination of NOM (Korshin et al., 1999; Leenheer et al., 2001; Fabbricino and Korshin, 2004). The behaviour of NOM fluorescence intensity during chlorination has been found to be non-monotonic, with the fluorescence intensity increasing in some NOM samples while decreasing in others (Fabbricino and Korshin, 2004).

Many different models for THMs formation during drinking water treatment have been developed. Many of these models were developed using multiple linear regression methods involving several water quality parameters as independent variables and formed THM concentrations as the predicted variables (Amy et al., 1987; Rodrıguez et al., 2003). Development of models for THMs aims to improve our understanding of the factors that contribute to the formation of THMs during water treatment and thus provides a decision support tool. If the developed models show that the THMs levels in the finished drinking water are likely to exceed the permissible maximum contaminant levels then water utilities could consider changing the primary disinfectant from chlorine to alternative disinfectants such as ozone, chlorine dioxide or chloramines. Models for prediction of THMs formation can also be used to optimise treatment processes for removal of precursor materials such as DOC.

DOC is widely used as an indicator of THM precursor material in drinking water. In order to minimise the formation of THMs during water treatment and limit the levels of THMs in treated water as well as to minimize chlorine decay in the distribution system, the removal of DOC prior to chlorine disinfection should be optimized to maintain low DOC concentrations in the treated water. Some of the methods used to optimise DOC removal include the standard jar test (Marhaba and Pipada, 2000; Vilge-Ritter et al., 1999, Yan et al., 2008), ultraviolet absorbance (UVA) measurements (Goslan et al., 2006; Gregor et al., 1997), zeta potential measurements (Sharp et al., 2006) and fluorescence measurements (Bieroza et al., 2009; Goslan et al., 2004; Marhaba et al., 2000). The efficiency of drinking water treatment is influenced by the amount and character of NOM present in water. Consequently, many water treatment utilities monitor NOM in their source waters, typically using bulk water quality parameters such as DOC concentration and UVA_{254}. In order to optimize treatment processes for the removal of NOM, a better understanding of its quantity as well as character is required.

Many investigations involving development of models for formation of THMs in drinking water treatment have used laboratory generated data but only a few have used full-scale water treatment plant data. This study investigates the incorporation of fluorescence measurements, which have relatively low expense and high sensitivity and can be relatively cheaply installed for online measurements, to improve the monitoring of THM formation in water treatment. Florescence measurements were based on fluorescence excitation-emission matrices (F-EEMs), which involve the use of excitation-emission wavelength pairs to identify fluorophores (fluorescent NOM fractions) based on the location of fluorescence peaks on F-EEM contour plots (Coble, 1996). These peaks have been used to distinguish between humic-like NOM, with longer emission wavelengths (> 350 nm), and protein-like NOM, with shorter emission wavelengths ($\leq$ 350 nm). Predictive models are developed for the removal of NOM and formation of THMs after chlorine disinfection in a full-scale drinking water treatment plant (WTP) using several water quality parameters which were measured as part of a study to characterize NOM and it's removal during water treatment. Multivariate statistical methods using simple linear regression and stepwise multiple linear regression (MLR) were used to model and predict the concentrations of DOC and THMs in the treated water.

7.2   Materials and methods

7.2.1 Sampling

Water samples were collected from Choisy-le-Roi (CR) drinking water treatment plant of Syndicat des Eaux d'lle de France (SEDIF), which supplies drinking water to the suburbs of the city of Paris, France, between March 2008 and September 2009. The treatment comprises conventional treatment coupled with biofiltration, using ozonation followed by biological activated carbon (BAC) filtration, and chlorine disinfection. Figure 7.1 shows the treatment process scheme and the sampling points for CR treatment plant. The following water samples were collected monthly: (i) raw water; (ii) preozonated water; (iii) settled water; (iv) sand filtered water; (v) ozonated water; (vi) BAC filtered water; and (vii) finished water.

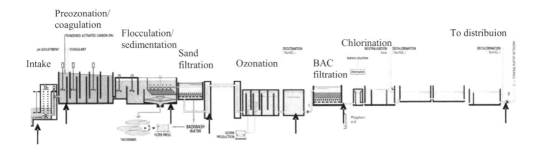

Figure 7.45 Treatment process scheme and sampling points for Choisy-le-Roi drinking water treatment plant.

The samples were collected in clean glass bottles and immediately filtered through 0.45 μm before being transported, within 24 hours, to the laboratory for analysis. The pre-filtered samples were stored at 5°C until required for analysis, which was normally done within one week of sampling. All the samples were analyzed for DOC concentration, UVA_{254} and F-EEMs. Besides the data generated from these analyses, water quality data indicative of flooding (e.g., turbidity) and algal (e.g., chlorophyll a, cell counts) events for raw-water samples as well as routine parameters measured (calcium, alkalinity, conductivity, pH, temperature, ozone doses and coagulation doses) were obtained from the treatment plant. Analyses for bromide and THMs (($CHCl_3$), $CHCl_2Br$, $CHBr_2Cl$ and $CHBr_3$) were performed by the Veolia Centre for Environmental Analysis, Saint-Maurice, France.

7.2.2 DOC and UVA_{254} measurements

DOC concentrations of all pre-filtered samples were determined by the catalytic combustion method using a Shimadzu TOC-V_{CPN} organic carbon analyzer. UVA_{254} of each sample was measured at room temperature (20±1°C) and ambient pH using a Shimadzu UV-2501PC UV-VIS scanning spectrophotometer. SUVA was determined by dividing the UVA_{254} by the corresponding DOC concentration.

7.2.3 Fluorescence Excitation Emission Matrices (F-EEM)

Fluorescence intensities for all samples were measured at ambient pH and room temperature ($20\pm1^{\circ}$C) using a FluoroMax-3 spectrofluorometer (Horiba Jobin Yvon). To account for fluorescence quenching resulting from relatively high DOC concentration in water samples, absorbance corrections have to be applied to fluorescence measurements. However, these time-consuming corrections are not necessary if the sample UVA_{254} absorbance is less than 0.05 cm^{-1} (Kubista et al., 1994) or if the DOC concentration of the sample is diluted to about 1 mg C/L prior to fluorescence measurement (Westerhoff et al., 2001). Since UVA_{254} absorbance was more than 0.05 cm^{-1} for nearly all raw water samples from the two water treatment plants, the prefiltered samples were diluted to a DOC concentration of 1 mg C/L using ultrapure water obtained from a Milli-Q$^{®}$ water purification system prior to fluorescence measurements.

F-EEMs were generated for each sample by scanning over excitation wavelengths between 240 and 450 nm at intervals of 10 nm and emission wavelengths between 290 and 500 nm at intervals of 2 nm.   An F-EEM of Milli-Q$^{®}$ water was obtained and this was subtracted from that of each sample in order to remove most of the water Raman scatter peaks. Since samples were diluted to a DOC concentration of 1 mg C/L prior to measurements, each blank subtracted F-EEM was multiplied by the respective dilution factor and Raman-normalized by dividing by the integrated area under the Raman scatter peak (excitation wavelength of 350 nm) of the corresponding Milli-Q$^{®}$ water, and the fluorescence intensities reported in Raman units (RU).

In this study, fluorescence parameters are given as maximum fluorescence intensities of peaks indentified from F-EEMs using the commonly applied approach of visual inspection of contour plots of EEMs (Coble, 1996). The identities and designations of the fluorescence intensity peaks selected based on the F-EEM contour plots of the samples collected are given in Table 3.1 (chapter 3). The table also shows the corresponding peaks identified by Coble (1996) and attributed to known fluorescing NOM compounds: tyrosine-like, tryptophan-like and humic-like NOM.

7.2.4 Data analysis and model development

The objective of the multiple linear regression analysis was to develop predictive models for DOC and THM concentrations and the required coagulant dose as a function of the measured water treatment plant variables. The data was analyzed using a spreadsheet program (Excel, Microsoft) and a statistical analysis program (SPSS). SPSS was used to run simple linear regression and stepwise multiple linear regression (MLR) models. Stepwise MLR involves classifying the predictor variables depending on their statistical significance and then searching for one variable at time that gives the best prediction of the outcome variable. In order to minimise the risks of making a Type II error (that is, missing a predictor variable that actually predicts the outcome), the backward method of stepwise multiple linear regression was used.

7.3   Results and discussion

7.3.1 Physico-chemical characteristics of source and finished water NOM

A total of 18 samples were collected from each of the sampling points of Choisy-le-Roi water treatment plant over the 18-month period. Results of some of the water quality parameters for source and finished waters are shown in Table 7.1. The DOC concentrations of the source and finished waters ranged between 2.0 mg C/L and 4.0 mg C/L, and 1.0 mg C/L and 2.2 mg C/L, respectively. The SUVA values for the source and finished waters varied between 1.7 and 3.7 L/mg-m, and 0.2 and 2.0 L/mg-m, respectively. NOM with SUVA values less than 3 L/mg-m is composed largely of non-humic organic matter and is relatively hydrophilic, less aromatic and of lower molecular weight compared to waters with higher SUVA values (Edzwald, 1993). For such waters, the DOC concentration has a small effect on coagulant doses and relatively low DOC removals by coagulation are expected (Edzwald, 1993). The mean DOC removal was 0.9 mg C/L by the coagulation-filtration process, representing a removal efficiency of 34%, and 0.3 mg C/L by BAC filtration, representing a removal efficiency of 17%.

Table 7.1 Data for a selection of water quality parameters for source and finished waters.

	Source water		Finished water	
Parameter	Mean + std.	Range	Mean + std.	Range
pH	8.0 ± 0.2	$7.65-8.19$	7.5 ± 0.2	$7.2-7.8$
Conductivity (μS/cm)	487 ± 43	$421-543$	519 ± 40	$453-572$
Temperature (°C)	17.4 ± 4.8	$8.4-26$	16.9 ± 5.0	$8.4-25$
Turbidity (FNU)	9 ± 6	$2-67$		
Alkalinity (mg/L CaCO3)	181 ± 17	$150-206$	178 ± 15	$144-207$
Bromide (mg/l)	0.05 ± 0.01	$0.03-0.06$		
Total algae (n/ml)	495 ± 1810	$0-7,957$		
TOC (mg C/L)	2.8 ± 0.7	$2.0-4.8$		
DOC (mg C/L)	2.6 ± 0.5	$2.0-4.0$	1.6 ± 0.3	$1.1-1.9$
UVA$_{254}$ (1/cm)	0.064 ± 0.019	$0.041-0.125$	0.016 ± 0.006	$0.003-0.029$
SUVA (L/mg-m)	2.52 ± 0.52	$1.69-3.67$	1.03 ± 0.39	$0.19-1.93$

7.3.2 Predictions using simple regression and stepwise multiple linear regression

Table 7.2 shows the variables measured during the 18-month period and used as independent variables to predict the source and finished water DOC concentrations, DOC removals, coagulant doses and total THMs concentration of the finished water. It also gives the means (± standard deviation) and minimum and maximum values of operational data such as coagulant and ozone doses, source water quality parameters such as pH, conductivity, turbidity, total alkalinity, UV$_{254}$, total algae counts, DOC concentration, bromide concentration and maximum fluorescence intensities of peaks B, T, M and C, as well as the bromate and total THMs concentrations of the finished water. The objective of using multiple linear regression analysis was to develop predictive models for finished water DOC and total THMs concentrations and the required coagulant dosage as functions of the measured water treatment plant variables. Backward stepwise multiple linear regression method, in which processing starts by including all the predictors in the model and then calculating the contributions of each predictor based on the significance value of its t-test, was used with SPSS. In this method, if the significance value of a predictor meets a removal criterion, the predictor is removed from the model and the model is re-estimated for the remaining predictors. After a regression model that best predicts the outcome has been selected, it needs to be assessed, using cross-validation, how accurately it can predict the outcome for other samples outside the training data set. To be able to perform a reliable cross-validation, which may involve data splitting techniques, the data set used to develop the regression model should contain a sufficient number of samples. However, because of limitations of time and financial resources, it is often not practical to collect sufficient data to perform a cross-validation. Under these circumstances, cross-validation could be performed using the adjusted R^2 value, which gives an indication of the loss of predictive power. In this study, several parameters were measured monthly over the 18-month period and the quantity of data generated would not allow cross-validation using data splitting. As such, the adjusted R^2 values were used to assess the predictive accuracy of the regression models and all the reported R^2 are the adjusted values.

Table 7.46 Input variables (variables 1-14) and predicted variables (variables 15-21) measured in the treatment plant.

Variable	Parameter	Mean ± std.	Range
1	pH	8.0 ± 0.2	7.7−8.2
2	Conductivity (µS/cm)	487 ± 43	412−543
3	Temperature (°C)	17.4 ± 4.8	8.4−25.6
4	Turbidity (NTU)	9 ± 16	2−67
5	Coagulant dose (mg Al/L)	22 ± 13	14−66
6	Ozone dose (mg O_3/L)	1.5 ± 0.4	1.0−2.5
7	Total Alkalinity (mg/L $CaCO_3$)	181 ± 17	150−206
8	Total algae (n/ml)	495 ± 1810	0−7957
9	UVA_{254} (1/cm)	0.064 ± 0.019	0.041−0.125
10	Peak B fluorescence intensity (R.U)	0.183 ± 0.034	0.113−0.239
11	Peak T fluorescence intensity (R.U)	0.258 ± 0.040	0.205−0.358
12	Peak M fluorescence intensity (R.U)	0.480 ± 0.097	0.363−0.683
13	Peak C fluorescence intensity (R.U)	0.424 ± 0.088	0.312−0.612
14	Bromide (mg/l)	0.05 ± 0.01	0.03−0.06
15	DOC (mg/l)	1.6 ± 0.3	1.1−1.9
16	Bromate (ug/l)	4.82 ± 2.17	2.30−9.30
17	THMs (ug/l)	8.79 ± 2.42	5.70−16.00

Simple linear regression could be applied to predict the source water DOC concentration using turbidity ($r^2 = 0.49$, $p < 0.001$) or UVA_{254} ($r^2 = 0.52$, $p < 0.001$). Results of analyses of water samples from two drinking water treatment plants showed stronger correlations ($r^2 > 0.90$) between DOC concentrations and UVA_{254} (Chowdhury and Champagne, 2008). UVA_{254} is commonly used as a surrogate for monitoring DOC concentrations and THM precursors in drinking water plants (Edzwald et al., 1985). However, an important limitation of UVA_{254} measurements is that they are effective mainly for unsaturated organic carbon fractions and not the entire organic matter spectrum. This study investigated whether the use of other easily and inexpensively measured parameters such as fluorescence measurements could be used to improve the prediction of the source water DOC. The ability to predict source water DOC concentration improved slightly ($r^2 = 0.58$) using a multiple linear regression relationship that included only temperature, conductivity and turbidity (Model 1,

Table 7.3). No reliable relationship could be found when fluorescence was included as one of the predictor variables.

Table 7.47 Predictions using simple and stepwise multiple linear regressions.

Model	Dependent variable	Variables in the model with *R*-squared value in []	Regression equation	*R*-squared (*N*)
1	Source water DOC (mg C/L)	conductivity (µS/cm) [0.01], temperature (°C) [0.04], turbidity (NTU) [0.49]	DOC_{source} = 0.005 cond + 0.052 temp + 0.037 turbidity – 1.137	0.58 (18)
2	Finished water DOC (mg C/L)	peak C fluorescence intensity (R.U) [0.24],	$DOC_{finished}$ = 3.164 peak C + 0.03	0.19 (17)
3	Finished water DOC (mg C/L)	component C1 fluorescence intensity (R.U.) [0.01], component C4 fluorescence intensity (R.U.) [0.07],	$DOC_{finished}$ = 3.70 comp C1 + 3.931 comp C4 – 0.799	0.30 (17)
4	DOC removal (%)	source water DOC (mg C/L) [0.28], conductivity (µS/cm) [0.22], coagulation dose (mg Al/L)) [0.22], total alkalinity (mg/L CaCO3) [0.40],	$DOC_{removal}$ = 14.393 DOC_{raw} – 0.102 cond – 0.269 coag – 0.584 alkal + 64.99	0.56 (15)
5	DOC removal (%)	coagulation dose (mg Al/L)) [0.22], total alkalinity (mg/L CaCO3) [0.40], peak M fluorescence intensity (R.U) [0.04],	$DOC_{removal}$ = 0.206 coag – 0.325 alkal +1 0.458 peak M + 88.764	0.37 (15)
6	Coagulation dose	temp (°C) [0.09], turbidity (NTU) [0.77], total alkalinity (mg/L CaCO3) [0.09], UVA (1/cm) [0.77], Peak T fluorescence intensity (R.U.) [0.00]	Coag_dose = 0.976 temp + 0.721 turb + 0.347 alkal + 287.03 UVA + 116.0 peak T –112.8	0.91 (17)
7	Total THMs (µg/L)	temp (°C) [0.49], turbidity (NTU) [0.08], ozone dose (mg O3/L) [0.18], UVA (1/cm) [0.06], peak T fluorescence intensity (R.U.) [0.16], peak M fluorescence intensity (R.U) [0.00],	THMs = 0.571 temp + 0.195 turb – 5.035 ozone – 147.387 UVA – 21.454 peak T + 30.582 peak M + 5.076	0.88 (16)
8	Total THMs (µg/L)	temp (°C) [0.46], turbidity (NTU) [0.05], ozone dose (mg O3/L) [0.13], UVA (1/cm) [0.03], component 1 fluorescence intensity (R.U.) [0.13], component 2 fluorescence intensity (R.U) [0.01],	THMs = 0.501 temp + 0.168 turb – 4.793 ozone – 122.597 UVA + 5.252 comp 1 + 18.960 comp2 + 3.321	0.74 (15)

Of the predictor variables used in simple linear regression to predict the finished water DOC concentration, only the source water DOC concentration (r^2 = 0.19), the ozone dose (r^2 = 0.20) and the peak C humic-like fluorescence (r^2 = 0.24) showed statistically significant correlations ($p < 0.05$). These relationships are much weaker than for source water DOC concentrations. No statistically significant relationship could be found using stepwise multiple linear regression except the one involving humic-like peak C fluorescence as the only predictor ($r^2 = 0.19$, $p < 0.05$). This relationship (Model 2, Table 7.3) is essentially a simple regression relationship. It shows that only about 20% of the variation in finished water DOC could be accounted for by humic-like peak C fluorescence, which is representative of relatively larger molecular weight and more hydrophobic humic NOM. Thus, based on the parameters measured and the data collected, the finished water DOC concentration could not be accurately predicted. The study also investigated whether the exclusive use of fluorescence characteristics of NOM in the source water to predict finished water DOC concentrations could be improved using a different set of fluorescence characteristics. This involved the use of the maximum fluorescence intensities of the five NOM fractions (components C1, C2, C3, C4 and C5) identified using PARAFAC analysis of F-EEMs for the same set of water samples (chapter 5) in stepwise multiple linear regression to predict finished water DOC concentration. Of the five components, only humic-like component C2 showed a statistically significant ($r^2 = 0.21$, $p < 0.05$) bivariate correlation with finished water DOC concentration. This correlation is comparable to the similar relationship with humic-like peak C ($r^2 = 0.24$) described earlier. Model 3 (Table 7.3) gives the multiple linear regression equation ($r^2 = 0.30$, $p < 0.05$) which includes only two of the five components, component C1 and component C4. Whereas the use of PARAFAC components slightly improves prediction of finished water DOC concentration using fluorescence measurements, the accuracy is still too low to be used in practice for optimization of drinking water treatment in terms of finished water DOC.

Multiple linear regression equations for the removal of DOC (%) were generated with and without source water DOC concentration as one of the predictor variables (Table 7.3, Models 4 and 5, respectively). The source water DOC concentration was left out in one of the regressions in order to investigate whether other routinely measured water quality parameters, which are more easily and more inexpensively measured online, could be used to predict DOC removal efficiency. If this were possible then it could provide water treatment operators with tools to optimise DOC removal in real-time due to sudden changes in source water qualities. The percentage DOC removal could be predicted ($r^2 = 0.56$, $p < 0.05$) using source DOC concentration, coagulation dose, conductivity and total alkalinity (Table 7.3, Model 4). When source DOC concentration was not included in the regression analysis, the DOC removal (%) could be predicted, though less accurately ($r^2 = 0.37$, $p < 0.05$), using coagulation dose, total alkalinity and peak C fluorescence (Table 7.3, Model 5).

Using simple linear regression, the coagulation dose could be predicted by turbidity ($r^2 = 0.77$, $p < 0.001$), UVA_{254} ($r^2 = 0.77$, $p < 0.001$) or source water DOC concentration ($r^2 = 0.74$, $p < 0.001$). Backward stepwise multiple linear regressions were performed with and without source water DOC concentration as one of the predictor variables. The latter produced a regression involving temperature, UVA_{254}, total alkalinity, turbidity and protein-like peak T fluorescence (Table 7.3, Model 6). When source water DOC concentration is included as one of the independent variables, the final regression model produced (the model is not shown in Table 7.3) includes DOC concentration and all of the parameters in the model without source water DOC concentration. For both cases, the prediction was significantly better than for simple linear regression ($r^2 = 0.91$, $p < 0.001$, without DOC, and $r^2 = 0.94$, $p < 0.001$, with DOC). The two models have similar predictive powers and since the model without DOC concentration has one fewer independent variable, it is preferable as it would obviate the need for measuring an additional parameter (DOC concentration). Figure 7.2 illustrates graphically the predicted coagulant doses using the multiple linear model (Model 6) versus the applied (measured) coagulant doses. The diagonal lines represent perfect agreement between the predicted and measured values.

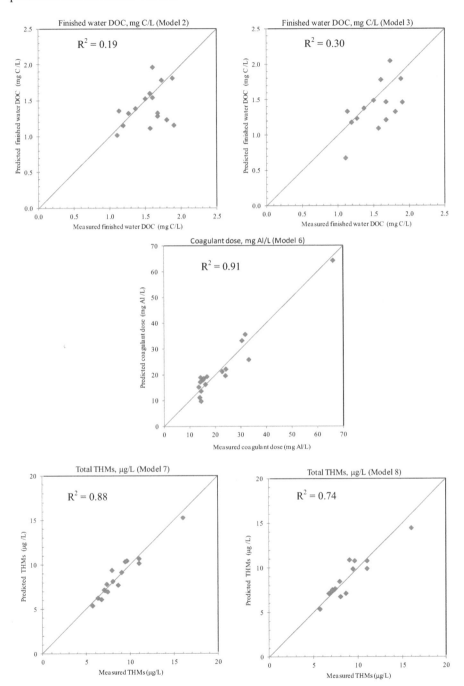

Figure 7.48 Comparison of measured data and model predictions using the same data set (the line shown depicts y = x).

Two sets of multiple linear regression analyses were performed for the prediction of total THMs in the finished water. For the first set, backward stepwise multilinear regression analysis was started with fourteen predictor variables (measured for source water, where applicable): conductivity, temperature, turbidity, coagulant dose, ozone dose, total alkalinity, total algae counts, source water DOC concentration, UVA_{254}, bromide concentration and fluorescence intensities of the four fluorescence peaks B, T, M and C. The resulting model (Table 7.3, Model 7) accurately predicted ($r^2 = 0.88$, $p < 0.001$) total THMs and it involved temperature, turbidity, ozone dose, UVA_{254} and fluorescence peaks T and M. The second set of multilinear regression analysis was performed in order to determine whether the model from the first set of regression analysis could be improved by replacing the fluorescence peaks B, T, M and C with PARAFAC components C1, C2, C3, C4 and C5. However, rather than starting with all the other independent parameters used in the first set of regression analysis, only the parameters in the final model (Model 7) and the five PARAFAC components, which replaced peaks T and M, were used. The final model developed (Table 7.3, Model 8), which involved temperature, turbidity, ozone dose, UVA_{254} and PARAFAC components C1 and C2 as predictor variables, had a slightly reduced prediction accuracy ($r^2 = 0.74$, $p < 0.001$) for total THMs in the finished water.

Comparisons of predicted and measured values of total THMs for the two models are graphically illustrated in Figure 7.2. The measured values of total THMs used for the comparison are the same values used for the regression analysis. A more accurate comparison would be achieved if there were a sufficient number of samples in the data set to permit splitting of the data set into two, one for the regression analysis and the other for validation. However, the performance of the models could still be investigated in the future using data collected in a new sampling campaign.

7.4   Conclusions

Based on the regression analysis performed in this study, the following conclusions about source and finished DOC concentration, coagulation dose and total THMs concentrations of the finished water can be made:

•   The source water DOC concentration correlated moderately with turbidity ($r^2 = 0.49$, $p < 0.001$) and UVA_{254} ($r^2 = 0.52$, $p < 0.001$) and its prediction improved slightly ($r^2 = 0.58$) using a multiple linear regression relationship that included temperature, conductivity and turbidity.

•   When fluorescence intensity was included as one of the predictor variables, no reliable multiple linear regression relationship for source water DOC concentration could be found.

•   Of the predictor variables used in simple linear regression to predict the finished water DOC concentration, only the source water DOC concentration ($r^2 = 0.19$), the ozone dose ($r^2 = 0.20$) and the peak C humic-like fluorescence ($r^2 = 0.24$) showed statistically significant correlations ($p < 0.05$).

•   Of the five fluorescence PARAFAC components, only the humic-like component C2 showed a statistically significant ($r^2 = 0.21$, $p < 0.05$) bivariate correlation with finished water DOC concentration. A multiple linear regression equation which includes two of the five components, component C1 and component C4, only slightly improved the prediction ($r^2 =$

0.30, $p < 0.05$) of finished water DOC concentration. Whereas the use of PARAFAC components slightly improves prediction of finished water DOC concentration using fluorescence measurements, the accuracy is still too low to be used in practice for optimization of finished water DOC.

•   The coagulation dose correlated well with turbidity ($r^2 = 0.77$, $p < 0.001$), UVA$_{254}$ ($r^2 = 0.77$, $p < 0.001$) or source water DOC concentration ($r^2 = 0.74$, $p < 0.001$). Prediction of coagulation dose was improved ($r^2 = 0.91$, $p < 0.001$) using multiple linear regressions involving temperature, UVA$_{254}$, total alkalinity, turbidity and protein-like peak T fluorescence. These parameters can be easily measured online and thus provide a practical and cost effective automatic adjustment of the applied coagulant dose.

•   The total THMs concentration of the finished water was accurately predicted ($r^2 = 0.88$, $p < 0.001$) using temperature, turbidity, ozone dose, UVA$_{254}$ and fluorescence peaks T and M. However, when fluorescence peaks T and M were replaced with the five PARAFAC components, the resulting model, which involved temperature, turbidity, ozone dose, UVA$_{254}$ and PARAFAC components C1 and C2, had a slightly reduced prediction accuracy ($r^2 = 0.74$, $p < 0.001$) for total THMs in the finished water.

•   Predictive modelling provides an alternative to complex analytical methods for determining the concentrations of THMs in finished drinking water, which could be more expensive and time consuming. Whereas the models for predictions of THMs concentrations show good predictability for the specific treatment plant investigated, they cannot reliably be globally applied to other water utilities without further research. The reliability of the models could be further investigated through the collection of more data which would permit cross-validation of the models using data splitting methods or by using the newly collected as the validation data set and the old data as the model development (training) data set.

7.5   References

Amy, G.L., Chadik, P.A. and Chowdhury, Z.K. 1987 Developing models for predcting THM formation potential and kinetics. *J. Am. Water Works Assoc.* 79, 89-97.

Ates, N., Kitis, M. and Yetis, U. 2007 Formation of chlorination by-products in waters with low SUVA—correlations with SUVA and differential UV spectroscopy. *Water Res.* 41, 4139-4148.

Bieroza, M., Baker, A. and Bridgeman, J. 2009 Relating freshwater organic matter fluorescence to organic carbon removal efficiency in drinking water treatment. *Sci. Total Environ.* 407(5), 1765-1774.

Chow, C.W.K., van Leeuwen, J.A., Drikas, M., Fabris, R., Spark, K.M. and Page, D.W. 1999 The impact of the character of natural organic matter in conventional treatment with alum. *Water Sci. Technol.* 40(9), 97-104.

Chowdhury, S., Champagne, P. and T., H. 2007 Fuzzy approach for selection of drinking water disinfectants. *J. Water Supply Res. Technol. AQUA* 56(2), 75-93.

Chowdhury, S. and Champagne, P. 2008 An investigation on parameters for modeling THMs formations. *Global NEST Journal* 10(1), 80-91.

Clark, R.M., Adams, J.Q., Sethi, V. and Sivaganesan, M. 1998 Control of microbial contaminants and disinfection by-products for drinking water in the US: cost and performance. *J. Water Supply Res. Technol. AQUA* 47(6), 255-265.

Coble, P.G. 1996 Characterization of marine and terrestrial DOM in seawater using excitation-emission matrix spectroscopy. *Marine Chemistry* 51, 325-346.

Edzwald, J.K., Becker, W.C. and Wattier, K.L. 1985 Surrogate parameter for monitoring organic matter and THM precursors. *J. Am. Water Works Assoc.* 77, 122-132.

Edzwald, J.K. 1993 Coagulation in drinking water treatment: particles, organics and coagulants. . *Water Sci. Technol.* 27(11), 21-35.

Fabbricino, M. and Korshin, G.V. 2004 Probing the mechanisms of NOM chlorination using fluorescence: formation of disinfection by-products in Alento River water. *Water Science and Technology: Water Supply* 4(4), 227–233.

Goslan, E.H., Voros, S., Banks, J., Wilson, D., Hillis, P., Campbell, A.T. and Parsons, S.A. 2004 A model for predicting dissolved organic carbon distribution in a reservoir water using fluorescence spectroscopy. *Water Res.* 38, 783-791.

Goslan, E.H., Gurses, F., Banks, J. and Parsons, S.A. 2006 An investigation into reservoir NOM reduction by UV photolysis and advanced oxidation processes. *Chemosphere* 65 1113–1119.

Gregor, J.E., Nokes, C.J. and Fenton, E. 1997 Optimising natural organic matter removal from low turbidity waters by controlled pH adjustment of aluminium coagulation. *Water Res.* 31(12), 2949-2958.

Korshin, G.V., Kumke, M.U., Li, C.W. and Frimmel, F.H. 1999 Influence of chlorination on chromophores and fluorophores in humic substances. *Environ. Sci. Technol.* 33(8), 1207-1212.

Korshin, G.V., Wu, W.W., Benjamin, M.M. and Hemingway, O. 2002 Correlations between differential absorbance and the formation of individual DBPs. *Water Res.* 36(13), 3273-3282.

Kubista, M., Sjoback, R., Eriksson, S. and Albinsson, B. 1994 Experimental Correction for the Inner-Filter Effect in Fluorescence-Spectra. *Analyst* 119(3), 417-419.

Leenheer, J.A., Rostad, C.E., Barber, L.B., Schroeder, R.A., Anders, R. and Davisson, M.L. 2001 Nature and Chlorine Reactivity of Organic Constituents from Reclaimed Water in Groundwater, Los Angeles County, California. *Environ. Sci. Technol.* 35(19), 3869-3876.

Marhaba, T.F. and Pipada, N.S. 2000 Coagulation: Effectiveness in removing dissolved organic matter fractions. *Environmental Engineering Science* 17(2), 107-115.

Marhaba, T.F., Van, D. and Lippincott, R.L. 2000 Rapid identification of dissolved organic matter fractions in water by spectral fluorescence signatures. *Water Res.* 34(14), 3543-3550.

Peters, C.J., Young, R.J. and Perry, R. 1980 Factors influencing the formation of haloforms in the chlorination of humic materials. *Environ. Sci. Technol.* 14(11), 1391-1395.

Platikanova, S., Puiga, X., Martı´nb, J. and Taulera, R. 2007 Chemometric modeling and prediction of trihalomethane formation in Barcelona's water works plant. *Water Res.* 41, 3394-3406.

Reckhow, D.A., Singer, P.C. and Malcolm, R.L. 1990a Chlorination of Humic Materials - by-Product Formation and Chemical Interpretations. *Environ. Sci. Technol.* 24(11), 1655-1664.

Reckhow, D.A., Singer, P.C. and Malcolm, R.L. 1990b Chlorination of humic materials: byproduct formation and chemical interpretations. *Environ. Sci. Technol.* 24(11), 1655-1664.

Richardson, S.D., Simmons, J.E. and Rice, G. 2002 Peer Reviewed: Disinfection Byproducts: The Next Generation. *Environ. Sci. Technol.* 36(9), 198A-205A.

Rodrıguez, M.J., Vunette, Y., Serodes, J.B. and Bouchard, C. 2003 Trihalomethanes in drinking water of greater Quebec region (Canada): occurrence, variations and modelling. *Environ. Monit. Assess.* 89(1), 69-93.

Sharp, E.L., Parsons, S.A. and Jefferson, B. 2004 The effects of changing NOM composition and characteristics on coagulation performance, optimisation and control. *Water Science and Technology: Water Supply* 4(4), 95-102.

Sharp, E.L., Parsons, S.A. and Jefferson, B. 2006 Coagulation of NOM: linking character to treatment. *Water Sci. Technol.* 53(7), 67-76.

Vilge-Ritter, A., Masion, A., Boulange, T., Rybacki, D. and Bottero, J.Y. 1999 Removal of natural organic matter by coagulation-flocculation: A pyrolysis-GC-MS study. *Environ. Sci. & Technol.* 33(17), 3027-3032.

Westerhoff, P., Debroux, J., Amy, G.L., Gatel, D., Mary, V. and Cavard, J. 2000 Applying DBP models to full-scale plants. *J. Am. Water Works Assoc.* 92(3), 89-102.

Westerhoff, P., Chen, W. and Esparza, M. 2001 Organic Compounds in the Environment Fluorescence Analysis of a Standard Fulvic Acid and Tertiary Treated Wastewater. *J. Environ. Qual.* 30, 2037–2046.

Yan, M., Wang, D., Qu, J., Ni, J. and Chow, C.W.K. 2008 Enhanced coagulation for high alkalinity and micro-polluted water: The third way through coagulant optimization. *Water Res.* 42, 2278 – 22

Chapter 8

SUMMARY AND CONCLUSIONS

8.1   Characterization of natural organic matter (NOM) in drinking water treatment processes and trains

Over the last 10–20 years, increasing natural organic matter (NOM) concentration levels in water sources have been observed in many countries. In addition to the trend towards increasing NOM concentration, the character of NOM can vary with source and time (season). The great seasonal variability and the trend towards elevated NOM concentration levels impose challenges to the water industry and water treatment facilities in terms of operational optimization and proper process control. By systematic characterization, the problematic NOM fractions can be targeted for removal and transformation. Therefore, proper characterization of the NOM in raw water or after different treatment steps would be an important basis for the selection of water treatment processes, monitoring of the performance of different treatment steps, and assessing distribution system water quality.

NOM in general significantly influences water treatment processes such as coagulation, oxidation, adsorption, and membrane filtration. In addition to aesthetic problems such as colour, taste and odour, NOM also contributes to the fouling of membranes, serves as precursor for the formation of disinfection by-products (DBPs) of health concern during oxidation processes and increases the exhaustion and usage rate of activation carbon. Furthermore, the biodegradable fraction of NOM may promote microbial growth in water distribution networks. Thus, in order to minimise these undesirable effects, it is essential to limit the concentration of NOM in the treated water. However, the efficiency of drinking water treatment is affected by both the amount and composition of NOM. Therefore, a better understanding of the physical and chemical properties of the various components of NOM would contribute greatly towards optimization of the design and operation of drinking water treatment processes.

Because of its complexity, the structure and fate of NOM in drinking water treatment (individual processes and process trains) are still not fully understood. Because it may contain thousands of different chemical constituents, it is not practical to characterize NOM on the basis of individual compounds. It is more feasible and the general practice to characterize NOM according to chemical groups or fractions having similar properties. These groups are commonly isolated by methods which involve concentration and fractionation of bulk NOM. Whereas these methods provide valuable insight into the nature of NOM from diverse aquatic environments, they are often laborious, time consuming and may involve extensive pre-treatment of samples which could modify the NOM character. They are also difficult to install for online measurement and are not commonly used for monitoring of NOM in drinking water treatment plants.

Analytical techniques that can be used to characterize bulk NOM without fractionation and pre-concentration and with minimal sample preparation are becoming increasingly popular. Non-destructive spectroscopic measurements require small sample volumes, are simple in practical application and do not require extensive sample preparation. This research aims at improving our understanding of the character and fate of NOM during different drinking water treatment processes using multiple NOM characterisation tools like F-EEM, SEC with UV and DOC detectors (SEC-OCD) and other bulk NOM water qualities such as UVA_{254}, SUVA and DOC. These complementary techniques could provide information on the fate of NOM fractions that negatively impact treatment efficiency, promote biological re-growth in water distribution systems and/or provide precursors for DBPs in systems that use oxidation/disinfection processes.

8.2 Characterization and influence of natural organic matter (NOM) in drinking water treatment

The presence of NOM in water significantly impacts different drinking water treatment processes as well as water quality in the distribution system, leading to operational problems and increased cost of water treatment. The study reviewed different NOM removal processes and the ways in which NOM affects drinking water quality and the performance of water treatment processes. The removal of NOM during drinking water treatment depends highly on the characteristics of the NOM present (e.g., molecular weight distribution (MWD), carboxylic acidity, and humic substances content), its concentration and the removal methods applied. High molecular weight (HMW) NOM is more amenable to removal than low molecular weight (LMW) NOM, particularly the fraction with an MW of 500 Dalton (Da). NOM components with the highest carboxylic functionality and hence the highest charge density are generally more difficult to remove by conventional treatment. Several water treatment methods have been used to remove NOM during drinking water treatment with varying degrees of success. A review of different tools for quantification and characterisation of NOM in drinking water treatment was carried out. Comparative analysis of different NOM characterization methods has demonstrated that there is no single method which can fully reveal NOM characteristics that are important for water treatment practice. The use of combinations of different methods would, therefore, be required for proper analysis of the fate of different fractions of NOM during different treatment processes. In situations where high skills and costly instruments are unavailable, a basic approach of tracking DOC and SUVA changes along the treatment process train could be used to understand the removal of NOM. High performance size exclusion chromatography coupled with UV/Vis, fluorescence, light scattering and sensitive dissolved organic carbon detection techniques could be used to obtain information on molecular absorbance, size distribution, molar mass and NOM reactivity. Information on biodegradability of NOM can be obtained using bioassays to determine the concentration of biodegradable dissolved organic carbon (BDOC) or assimilable organic carbon (AOC).

8.3 Characterizing natural organic matter in drinking water: From source to tap

Natural organic matter (NOM) in water samples from two drinking water treatment trains with distinct water quality, and from a common distribution network with no chlorine residual, was characterized and the relation between biological stability of drinking water and NOM was investigated. NOM was characterised using fluorescence excitation–emission matrices (F-EEMs), size exclusion chromatography with organic carbon detection (SEC-OCD) and assimilable organic carbon (AOC). The treatment train with higher concentrations of humic substances produced more AOC after ozonation. NOM fractions determined by SEC-OCD, as well as AOC fractions, NOX and P17, were significantly lower for finished water of one of the treatment trains. F-EEM analysis showed a significantly lower humic-like fluorescence for that plant, but no significant differences for the tyrosine- and tryptophan-like fluorescence. For all of the SEC-OCD NOM fractions, the concentrations in the distribution system were not significantly different than in the finished waters. For the common distribution network, distribution points supplied with finished water containing higher AOC and humic substances concentrations had higher concentrations of ATP and *Aeromonas* sp. The number of aeromonads in the distribution network was significantly higher than in the finished waters, whereas the total ATP level remained constant, indicating no overall bacterial growth.

8.4   Tracking NOM in a drinking water treatment plant using F-EEM and PARAFAC

The use of F-EEMs and PARAFAC to characterize NOM in drinking water treatment and the relationship between the extracted PARAFAC components and the corresponding SEC-OCD fractions was investigated. A seven component PARAFAC model was developed and validated using 147 F-EEMs of water samples from two full-scale water treatment plants. The fluorescent components have spectral features similar to those previously extracted from F-EEMs of dissolved organic matter (DOM) from diverse aquatic environments. Five of these components are humic-like with a terrestrial, anthropogenic or marine origin, while two are protein-like with fluorescence spectra similar to those of tryptophan-like and tyrosine-like fluorophores. A correlation analysis was carried out for samples of one treatment plant between the maximum fluorescence intensities (F_{max}) of the seven PARAFAC components and NOM fractions (humics, building blocks, neutrals, biopolymers and low molecular weight acids) of the same sample obtained using SEC-OCD. There were significant correlations ($p < 0.01$) between sample DOC concentration, UVA_{254}, and F_{max} for the seven PARAFAC components and DOC concentrations of the SEC-OCD fractions. Three of the humic-like components showed slightly better predictions of DOC and humic fraction concentrations than did UVA_{254}. Tryptophan-like and tyrosine-like components correlated positively with the biopolymer fraction. These results demonstrate that fluorescent components extracted from F-EEMs using PARAFAC could be related to previously defined NOM fractions and that they could provide an alternative tool for evaluating the removal of NOM fractions of interest during water treatment.

8.5   Characterizing NOM and removal trends during drinking water treatment

Natural organic matter (NOM) is of concern in drinking water because it causes adverse aesthetic qualities such as taste, odour, and colour; impedes the performance of treatment processes; and decreases the effectiveness of oxidants and disinfectants while contributing to undesirable disinfectants by-products. The effective removal of NOM during drinking water treatment requires a good understanding of its character. Because of its heterogeneity, NOM characterization necessitates the use of multiple analytical techniques. In this study, NOM in water samples from two drinking water treatment trains was characterized using SEC-OCD and F-EEMs with PARAFAC. These characterization methods showed that the raw and treated waters were dominated by humic substances. Whereas the coagulation process for both plants may be optimized for the removal of bulk DOC, it is not likewise optimized for the removal of specific NOM fractions. A five component PARAFAC model was developed for the F-EEMs, three of which are humic-like, while two are protein-like. These PARAFAC components and the SEC-OCD fractions represented effective tools for the performance evaluation of the two water treatment plants in terms of the removal of NOM fractions.

8.6 Investigating the impact of water treatment on the fluorescence spectra of humic substances in surface and ground waters

This study investigates the effects of different water treatment processes for the removal of natural organic matter (NOM) in surface and ground waters on the fluorescence characteristics of the NOM. Online monitoring of dissolved organic carbon (DOC) concentration using fluorescence excitation-emission matrices (F-EEM) would generally be based on a fixed pair of excitation and emission wavelengths, such as for humic-like peak C or protein-like peak T. However, some treatment processes are known to result in a shift in the location of the fluorescence peaks and using the same pair of excitation emission wavelengths could potentially result in errors in the measurement of the maximum fluorescence intensity. This study focuses on the spectral shifts of peak C and investigates the amount of error in the fluorescence intensity maximum if the shift in the location of peak C is not taken into account. Raw and treated surface and ground water samples were analyzed for F-EEM and the shift in the fluorescence spectra as well as the percentage error of the fluorescence intensity maximum of peak C were determined. The samples were treated for NOM removal in coagulation jar tests, pilot plants and full-scale water treatment plants. Coagulation of surface and ground water with iron chloride and alum resulted in a shift in the emission wavelength of humic-like peak C of between 8 and 18 nm, and an error in the maximum fluorescence intensity ranging between 2% and 6% if the shift is not taken into account. There was no significant difference in the spectral shift of peak C or in the error in the maximum fluorescence intensity between coagulation alone and coagulation followed by ozonation of ground water. NOM removal with ion exchange (IEX) alone generally resulted in a higher shift in peak C and a higher percentage error in the maximum fluorescence intensity than with coagulation, biological activated carbon (BAC) filtration or a combination of treatments. The impact of IEX treatment on the error of maximum fluorescence intensity was higher for surface than for ground waters, likely due to differences in molecular weight distribution of surface and ground water NOM. The results demonstrate that for NOM removal treatments other than IEX, the errors in the maximum fluorescence intensity that would result from ignoring the fluorescence spectral shifts are generally low ($\leq 5\%$), and a fixed excitation emission wavelength pair for peak C could be used for online monitoring of NOM in water treatment plants. If IEX is included in the water treatment train, the resultant spectral shifts and fluorescence intensity errors should be ignored only if it is located at the start of the treatment train.

8.7 Modelling and prediction of the removal of NOM and formation of THMs in drinking water treatment

This study investigates the incorporation of fluorescence measurements, which have relatively low expense and high sensitivity and can be relatively cheaply installed for online measurements, to improve the monitoring of concentrations of dissolved organic carbon (DOC) and total trihalomethanes (THMs) in a drinking water treatment. Florescence measurements were based on fluorescence excitation-emission matrices (F-EEMs), which involve the use of excitation-emission wavelength pairs to identify fluorophores (fluorescent NOM fractions) based on the location of fluorescence peaks on F-EEM contour plots (Coble, 1996). Predictive models are developed for the removal of NOM and formation of THMs after chlorine disinfection in a full-scale drinking water treatment plant (WTP) using several water quality parameters which were measured as part of a study to characterize NOM and it's removal during water treatment. Statistical methods using simple linear regression and stepwise multiple linear regression (MLR) were used to model and predict the concentrations of DOC and THMs in the treated water.

The source water DOC concentration could be moderately predicted ($r^2 = 0.58$) using a multiple linear regression relationship that included temperature, conductivity and turbidity. Whereas the use of PARAFAC fluorescence components slightly improved the prediction of finished water DOC concentration, the prediction accuracy was generally low for both simple and multiple linear regressions. The applied coagulation dose could be predicted ($r^2 = 0.91$, $p < 0.001$) using multiple linear regressions involving temperature, UVA$_{254}$, total alkalinity, turbidity and protein-like peak T fluorescence. The total THMs concentration of the finished water could be predicted ($r^2 = 0.88$, $p < 0.001$) using temperature, turbidity, ozone dose, UVA$_{254}$ and fluorescence peak T and humic-like peak M. However, when fluorescence peaks T and M were replaced with the five PARAFAC components, the resulting model, which involved temperature, turbidity, ozone dose, UVA$_{254}$ and PARAFAC components C1 and C2, had a slightly reduced prediction accuracy ($r^2 = 0.74$, $p < 0.001$) for total THMs in the finished water.

Predictive modelling provides an alternative to relatively complex analytical methods for determining the concentrations of THMs in finished drinking water, which could be more expensive and time consuming. Whereas the models for predictions of THMs concentrations show good predictability for the specific treatment plant investigated, they cannot reliably be globally applied to other water utilities without further research.

8.8   Recommendations

This research aims at improving our understanding of the character and fate of NOM during different drinking water treatment processes using multiple NOM characterisation tools like F-EEM, SEC with UV and DOC detectors (SEC-OCD) and other bulk NOM water qualities such as UVA$_{254}$, SUVA and DOC. These complementary techniques could provide information on the fate of NOM fractions that negatively impact treatment efficiency, promote biological re-growth in water distribution systems and provide precursors for DBPs in systems that use oxidation/disinfection processes. Some recommendations for future research towards improving our understanding of NOM character are:

- This research has demonstrated that F-EEMs can be used with PARAFAC to decompose the fluorescent spectra of bulk NOM into fluorescence spectra of individual PARAFAC components. However, there is still lack of information regarding what these PARAFAC components actually represent. There is, therefore, a need for further research to try and identify these components and to determine whether they represent individual NOM fractions or groups of fractions having similar fluorescence characteristics. If a database of these PARAFAC components could be built and then related to known compounds, our understanding of a significant fraction of NOM would be enhanced.

- Whereas this research has shown that SEC-OCD is an effective tool for characterizing NOM in drinking water treatment in the evaluation of water treatment performance in terms of NOM removal, there is still more work that is required in the fractionation and detection of the low molecular weight acids (LMW acids). The SEC-OCD system used in this research uses an empirical approach for detection and quantification of LMW acids but the results are sometimes less than convincing. Further research could consider using chromatographic columns which specifically target the fractionation of the LMW organics.

List of abbreviations

AC	Activated carbon
AER	Anionic exchange resins
AMW	Apparent molecular weight
AOC	Assimilable organic carbon
AOM	Algal organic matter
ATP	Adenosine triphosphate
BAC	Biological activated carbon
BBs	Building blocks
BDOC	Biodegradable dissolved organic carbon
BF	Bank filtration
BPs	Biopolymers
CR	Choisy-le-Roi
DBPs	Disinfection by-products
DN1	Water distribution point 1
DN2	Water distribution point 2
DN3	Water distribution point 3
DOC	Dissolved organic carbon
DOM	Dissolved organic matter
DON	Dissolved organic nitrogen
EfOM	Effluent organic matter
F-EEMs	Fluorescence excitation emission matrices
FA	Fulvic acids
F_{max}	Maximum fluorescence intensity
FRI	Fluorescence regional integration
GAC	Granular activated carbon
HA	Humic acids
HMW	High molecular weight
HPLC	High performance liquid chromatographic
HPSEC	High performance size exclusion chromatography
HS	Humic substances
IEX	Ion exchange
LDN	Leiduin
LMW	Low molecular weight
LVN	Loenderveen
MIEX	Magnetic Ion Exchange Resins
MWCO	Molecular weight cut-off
MW	Molecular weight
NDMA	N-nitrosodimethylamine
NF	Nanofiltration
NM	Neuilly-sur-Marne
NOM	Natural organic matter
PAC	Powdered activated carbon
PARAFAC	Parallel factor analysis
PCA	Principal Component analysis
PLS	Partial least squares regression,
PRAM	Polarity rapid assessment method
RSF	Rapid sand filtration

RU	Raman units
SEC-OCD	Size exclusion chromatography with organic carbon detection
SEDIF	Syndicat des Eaux d'Ile de France
SPE	Solid-phase extraction
SSF	Slow sand filtration
SUVA	Specific UV absorbance, defined as ratio of UVA_{254} to DOC concentration
TOC	Total organic carbon
UF	Ultrafiltration
URI	UV absorbance ratio index
UV-Vis	UV and visible
UVA_{254}	UV absorbance at a wavelength of 254 nm
UVA	UV absorbance
UV	Ultraviolet
WPK	Weesperkarspel
ΔUVA	Differential UV absorbance

Samenvatting

De laatste 10-20 jaar is in meerdere landen een stijging van de natuurlijk organisch materiaal (NOM) concentratie in water bronnen waargenomen. Mogelijk oorzaken voor deze stijging zijn het broeikaseffect, bodem verzuring, ernstige droogte en vaker intensive regenval. Niet alleen de NOM concentratie, maar ook de samenstelling van het NOM varieert per bron en tijd (seizoen). De grote seizoen variatie en de trend naar verhoogde NOM concentraties zorgen voor uitdagingen voor de drinkwaterindustrie en de waterzuiveringsinstanties in termen van operationele optimalisatie and goede proces controle. Door systematische karakterisering, kunnen de problematische NOM fracties worden bepaald en gericht worden verwijderd of omgezet. Daarom zou een goede karakterisering van het NOM in het ruwe water of na verschillende zuiveringsstappen een belangrijke basis zijn voor de selectie van water zuiveringsprocessen, bewaking van de verschillende zuiveringsprocessen en het beoordelen van de waterkwaliteit in het distributiesysteem.

NOM is een heterogeen mengsel van natuurlijk voorkomende organische verbindingen die veelvuldig gevonden worden in grond- en oppervlaktewater. NOM is afkomstig uit levende en dode planten, dieren en micro-organismen en de afbraakproducten van dezen. NOM in zijn algemeenheid heeft een significante invloed op waterzuiveringsprocessen zoals coagulatie, oxidatie, adsorptie en membraan filtratie. Naast esthetische problemen zoals kleur, smaak en geur, draagt NOM ook bij aan de vervuiling van membranen, dient als precursor voor de vorming van desinfectie bijproducten (DBP) tijdens desinfectie / oxidatieprocessen wat gezondheidsrisico's met zich mee brengt en NOM verhoogt de verzadigingssnelheid en gebruikte hoeveelheid van actief kool. Bovendien kan de biologisch afbreekbare fractie van NOM microbiologische groei in waterleidingnetten bevorderen. De efficiëntie van de zuivering van drinkwater wordt beïnvloed door zowel de hoeveelheid als de samenstelling van het NOM. Daarom zou een beter begrip van de fysische en chemische eigenschappen van de verschillende componenten van NOM in grote mate bijdragen tot optimalisatie van het ontwerp en de werking van drinkwaterzuiveringsprocessen.

Het is mogelijk dat NOM bestaat uit duizenden verschillende chemische bestanddelen, daarom is het in de praktijk niet handig om NOM te karakteriseren op basis van afzonderlijke verbindingen. Het is meer haalbaar en de algemene praktijk om NOM ter karakteriseren op basis van chemische groepen met soortgelijke eigenschappen. Deze groepen worden gewoonlijk gescheiden door methodes die gebruik maken van concentratie en fractionering van bulk NOM. Deze werkwijzen zijn vaak arbeidsintensief, tijdrovend en voorbehandeling van de monsters kan aan de orde zijn, waardoor de NOM samenstelling kan veranderen. Deze methodes zijn ook moeilijk te gebruiken voor online metingen en ze worden niet vaak gebruikt voor het monitoren van NOM in drinkwater zuiveringsinstallaties.

Analytische technieken die kunnen worden gebruikt voor het karakteriseren van bulk NOM, zonder fractioneren en pre-concentratie en met minimaal monster voorbehandeling worden steeds populairder. Hoogwaardige gelpermeatiechromatografie (HPSEC) en fluorescentie excitatie-emissie matrix (F-EEM) spectroscopie worden steeds meer gebruikt voor het karakteriseren van NOM in drinkwater. Meer gedetailleerde informatie over NOM kan worden verkregen door het gebruik van F-EEM spectra en parallel factoranalyse (PARAFAC). PARAFAC is een statistische methode om meerdimensionale data ontleden.

Het doel van dit onderzoek was om bij te dragen aan een beter begrip van het karakter van NOM voor en na zuivering van verschillende drinkwaterbehandelingsprocessen alsook in het water distributienetwerk door gebruik te maken van verschillende NOM karakteristeringsmethodes zoals F-EEM, SEC met ultraviolet absorptie (UVA) en opgelost organisch koolstof (DOC) detectoren (SEC-OCD), en andere bulk NOM analyses zoals UVA bij 254 nm (UVA$_{254}$), specifieke UVA$_{254}$ (SUVA) en DOC. Deze complementaire technieken kunnen informatie verschaffen over NOM fracties die een negatieve invloed hebben op de zuiveringsefficiëntie, die biologische nagroei in het water distributie systemen bevorderen en NOM fracties die dienen als precursoren voor DBP in zuiveringen die gebruik maken van oxidatie / desinfectie processen. De verwachting is dat dit de optimalisatie van NOM verwijdering gedurende de waterzuivering mogelijk maakt op het gebied van kwantiteit en op specifieke NOM fracties die de besturing van de zuivering en de gezondheid beïnvloeden.

NOM in water monsters van twee drinkwaterzuiveringen, met verschillende water kwaliteit en met een gedeeld distributie netwerk zonder chloor, was gekarakteriseerd en de relatie tussen de biologische stabiliteit van het drinkwater en NOM was bepaald door assimileerbaar organisch koolstof (AOC) analyses. NOM was gekarakteriseerd door F-EEM, SEC-OCD and AOC. De zuivering met hogere concentraties humus zuren produceerde meer AOC na ozon. NOM fracties bepaald door SEC-OCD, alsook de AOC fracties NOX en P17 waren significant lager voor het behandelde water van één van de zuiveringen. F-EEM analyses lieten een significant lagere humus-achtige fluorescence zien voor deze zuivering, maar geen significate verschillen voor tyrosine- en tryptofaan-achtige fluorescence. Voor geen van de SEC-OCD fracties waren de concentraties in het distributienetwerk significant verschillend met het water in de reinwaterkelder. Punten in het gedeelde distributienetwerk die werden geleverd met water met een hogere AOC concentratie en humuszuren concentratie hadden een hogere concentratie van adenosine tri-fosfaat (ATP) en Aeromonas sp. Het aantal aeromonads in het distributienetwerk was significant hoger dan in het behandelde water, terwijl de totale ATP concentratie constant bleef. Dit geeft een indicatie dat er geen bacteriële groei plaatsvindt.

Het gebruik van F-EEMs en PARAFAC voor NOM karakterisatie in de drinkwater zuivering en de relatie tussen de bepaalde PARAFAC componenten en de bijbehorende SEC-OCD fracties is onderzocht. Een zeven componenten PARAFAC model was ontwikkeld en gevalideerd door gebruik te maken van 147 F-EEMs van water monsters van twee water zuiveringsinstallaties. Vijf van deze componenten zijn humus-achtig met een aardse, menselijke of marine oorsprong, terwijl twee componenten eiwit-achtig zijn met fluorescentie spectra vergelijkbaar met die van tryptofaan en tyrosine-achtige fluoroforen. Er is een correlatie analyse gedaan voor monsters van een zuivering tussen de maximale fluorescence intesitiet (F$_{max}$) van de zeven PARAFAC componenten en de NOM fracties van dezelfde monsters verkregen door het gebruik van SEC-OCD. De DOC concentraties, UVA$_{254}$ en F$_{max}$ in de monsters voor de zeven PARAFAC componenten correleerde significant (p<0.01) met de concentraties van de SEC-OCD fracties. Drie van de humus-achtige componenten lieten een iets betere voorspelling van DOC en humus fractie concentratie zien dan de UVA$_{254}$. Tryptofaan-achtige en tyrosine-achtige componenten correleerden positief met de biopolymeer fractie. Deze resultaten laten zien dat fluorescent componenten bepaald vanuit de F-EEMs door gebruik te maken van PARAFAC kunnen worden gerelateerd aan eerder gedefinieerde NOM fracties en dit kan een alternatieve methode bieden voor het evalueren van de verwijdering van specifieke NOM fracties tijdens waterzuivering.

NOM in watermonsters uit twee drinkwaterzuiveringen werd gekarakteriseerd met behulp van SEC-OCD en F-EEMs met PARAFAC. Deze karakterisering methoden laten zien dat het ruwe en behandelde water gedomineerd werd door humuszuren. De PARAFAC componenten en SEC-OCD fracties werden vervolgens gebruikt om de prestaties van de zuiveringen te beoordelen op basis van de verwijdering van de verschillende NOM fracties. De coagulatie is voor beide zuiveringen geoptimaliseerd voor de verwijdering van bulk DOC, dit betekent niet dat het ook geoptimaliseerd is voor de verwijdering van specifieke NOM fracties. Een vijf componenten PARAFAC model is ontwikkeld voor de F-EEMs, waarvan er drie humus-achtig en twee eiwit-achtige zijn. Deze PARAFAC componenten en de SEC-OCD fracties zijn nuttig gebleken als extra hulpmiddelen voor de evaluatie van de prestaties van de twee waterzuiveringsinstallaties voor de verwijdering van specifieke NOM fracties.

De invloed van verschillende waterbehandelingsprocessen voor de verwijdering van NOM uit oppervlaktewater en grondwater op de fluorescentie eigenschappen van de NOM is onderzocht. Het onderzoek gaat in op de fluorescentie spectrale verschuivingen van de humus-achtige piek (piek C) bij een excitatiegolflengte in het zichtbare gebied van 300-370 nm en een emissie golflengte tussen 400 en 500 nm, en onderzoekt de hoeveelheid fouten in de bepaling van het fluorescentie-intensiteit maximum, als er geen rekening gehouden wordt met de verschuiving in de locatie van de piek C. Coagulatie van oppervlakte- en grondwater met ijzerchloride en aluminium resulteerde in een verschuiving in de emissiegolflengte van de humus-achtige piek C van 8 tot 18 nm, en een fout in de maximale fluorescentie intensiteit tussen de 2% en 6% als er geen rekening gehouden wordt met de verschuiving. Er was geen significant verschil in de spectrale verschuiving van piek C of in de fout in de maximale fluorescentie intensiteit tussen alleen coagulatie en coagulatie gevolgd door ozonisatie van grondwater. NOM verwijdering met alleen ionenwisseling (IEX) leidt in het algemeen tot een hogere verschuiving van piek C en een hoger procentuele fout in de maximale fluorescentie intensiteit dan met coagulatie, biologische actieve kool (BAC) filtratie of een combinatie van deze behandelingen. Het effect van IEX behandeling op de fout in de maximale fluorescentie intensiteit was hoger voor oppervlakte water dan voor grondwater, waarschijnlijk als gevolg van verschillen in molecuulgewichtverdeling van het NOM in oppervlakte-en grondwater. De resultaten tonen aan dat voor zuiveringsprocessen die NOM verwijderen, anders dan IEX, de fouten in de maximale fluorescentie-intensiteit als gevolg van het negeren van de fluorescentie spectrale verschuivingen in het algemeen laag zijn($\leq 5\%$) en een vast paar van excitatie emissiegolflengte voor piek C kan worden gebruikt voor online monitoring van NOM in waterzuiveringsinstallaties.

Gebruik van F-EEMs voor een verbetering van de monitoring van de DOC concentraties en de totale trihalomethanen (THM) in drinkwaterzuivering is geëvalueerd. Voorspellende modellen zijn ontwikkeld voor de waterkwaliteit voor het verwijderen van NOM en de vorming van THM na desinfectie met chloor met behulp van verschillende gemeten parameters in een full-scale drinkwaterzuivering. Het gebruik van PARAFAC fluorescentie componenten zorgt voor een lichte verbetering in de voorspelling van de DOC concentratie in het reine water, maar de nauwkeurigheid van de voorspelling was over het algemeen laag voor zowel de eenvoudige lineaire als de meervoudige lineaire regressies. De toegepaste coagulatie dosis kan worden voorspeld ($r^2 = 0,91$, p $<0,001$) met meervoudige lineaire regressie met temperatuur, UVA254, totale alkaliniteit, troebelheid en tryptofaan-achtige fluorescentie (piek T). De totale concentratie van de THMs in het reine water was te voorspellen ($r^2 = 0,88$, p $<0,001$) met temperatuur, troebelheid, ozondosering, UVA_{254}, fluorescentiepiek T en humus-achtige piek (piek M) met een excitatie maximum van 310 nm en een emissiemaximum van 410 nm.

Dit onderzoek draagt bij aan onze kennis van het karakter van NOM en de impact van verschillende drinkwaterbehandelingsprocessen op zijn eigenschappen. Het toont het nut van het gebruik van meerdere NOM karakterisering methodes die kunnen worden gebruikt voor de selectie en de werking van de zuiveringsprocessen, bewaking van de prestaties van de verschillende behandelingsstappen, en de beoordeling van de waterkwaliteit in een waterdistributiesysteem.

List of publications

1. Baghoth, S.A., Sharma, S.K. and Amy, G.L. 2011 Tracking natural organic matter (NOM) in a drinking water treatment plant using fluorescence excitation-emission matrices and PARAFAC. *Water Res.* 45(2), 797-809.

2. Baghoth, S.A., Sharma, S.K., Guitard, M., Heim, V., Croue, J.P. and Amy, G.L. 2011 Removal of NOM-constituents as characterized by LC-OCD and F-EEM during drinking water treatment. *J. Water Supply. Res. Technol. AQUA* 60(7), 412-424.

3. Baghoth, S.A., Dignum, M., Grefte, A., Kroesbergen, J. and Amy, G.L., 2010 Characterization of NOM in a drinking water treatment process train with no disinfectant residual. *Water Sci. Technol. Water Supply* 9(4): 379-386.

4. Baghoth, S.A., Maeng, S.K., Salinas Rodríguez, S.G., Ronteltap, M., Sharma, S., Kennedy, M. and Amy, G.L. 2008 An urban water cycle perspective of natural organic matter (NOM): NOM in drinking water, wastewater effluent, storm water, and seawater. *Water Sci. Technol. Water Supply* 8(6), 701-707.

5. Baghoth, S.A., Dignum, M., Sharma, S.K. and Amy, G.L. Characterizing natural organic matter (NOM) in drinking water: from source to tap. Submitted to *J. Water Supply. Res. Technol. AQUA (under review)*.

6. Baghoth, S.A., Mosebolatan, K.O., Sharma, S.K. and Amy, G.L. Investigating the impact of water treatment on the fluorescence spectra of humic substances in surface and ground waters. Submitted to *Water Science and Technology*.

7. Baghoth, S.A., Sharma, S.K. and Amy, G.L. Modelling and prediction of the removal of NOM and formation of trihalomethanes in drinking water treatment. Submitted to *Water Science and Technology*.

8. Baghoth, S.A., Sharma, S.K. and Amy, G. Characterization and influence of natural organic matter (NOM) in drinking water treatment: A review. In preparation for submission to *Water Res.*

About the Author

Saeed Abdallah Baghoth was born in Kamuli, Uganda, on October 22, 1964. He attended secondary school at St. Mary's College, Kisubi, Uganda, from 1978 to 1984. He joined Makerere University, Kampala, Uganda, in 1984 and graduated with a Bachelor's degree in Engineering (Civil) with honours in 1989. He joined the family business and worked until 1994, when he travelled to Yemen and worked with Shibaam Engineering Co. as an assistant site engineer during the construction of a clay bricks and tiles factory in Aden. He returned to Uganda in 1995 and joined Kamuli Town council as the Town Engineer and worked until 2000. In October 2000, he was appointed the Water Engineer for Kamuli District Local Government where he managed the Rural Water and Sanitation Programme. In 2002, he was awarded a scholarship from the Netherlands Fellowship Programme (NFP) to study Sanitary Engineering at IHE Delft (now UNESCO-IHE). He did his MSc. research on fouling of ultra-filtration membranes and graduated with a MSc. degree with distinction in 2004. He then returned to Uganda to continue working with the Rural Water and Sanitation Programme in Kamuli until 2006, when he was again awarded a scholarship to start his PhD study at UNESCO-IHE Institute for Water Education/ Delft University of Technology. He is currently a senior water engineer with Kamuli district local government.

T - #0421 - 101024 - C180 - 246/174/10 - PB - 9781138000261 - Gloss Lamination